EARTHQUAKE ENGINEERING

Edited by
Judith Rosales

Delve Publishing

Earthquake Engineering

Edited by Judith Rosales

Edition 2018

Delve Publishing

Published by

Delve Publishing LLC

708 3rd Avenue, 6th Floor

New York

NY 10801

United States of America

www.delvepublishing.com

ISBN: 978-1-68095-734-1 (Hardback)

Printed in United States of America

Library of Congress Control Number: 2016920414

About the Editor

Judith Rosales

Prof. Judith is an Ecologist and Environmental Specialist with more than 30 years of academic and consulting experience. She holds a PhD in geography from the University of Birmingham UK and wrote the PhD and Master Academic Programs of Environmental Sciences for the Universidad de Guayana, Venezuela.

Table of Contents

2 TYPES OF EARTHQUAKE

3 CAUSES OF EARTHQUAKE

4 MEASUREMENT OF EARTHQUAKES

5 SEISMIC BASE ISOLATION

6 VIBRATION ISOLATION

7 EARTHQUAKE RESISTANT STRUCTURES

8 DISASTER MANAGEMENT

9 GEOTECHNICAL ENGINEERING

10 SEISMIC RETROFITTING

List of Abbreviations

AGP	Academy of Geo-Professionals
ARG	Architectural Resources Group
CBCP	Certified Business Continuity Professional
CEM	Certified Emergency Manager
CEUS	Central and Eastern United States
CW	Continuous Wave
DHS	Department of Homeland Security
DR	Damage Ratio
EERI	Earthquake Engineering Research Institute
EGS	Enhanced Geothermal Systems
EIT	Extreme Ultraviolet Imaging Telescope
EMIS	Emergency Management Information Systems
ESPI	Electronic Speckle Pattern Interferometry
ETAS	Epidemic-Type Aftershock Sequence model
ETS	Episodic Tremor and Slip
FACT	Field Assessment and Coordination Teams
FD	Friction Dampers
FDM	Frequency Division Multiplexing
FEMA	Federal Emergency Management Agency
FPB	Friction Pendulum Bearing
FPS	Friction Pendulum System
FVD	Fluid viscous dampers
HDR	Hot Dry Rock

IAEM	International Association of Emergency Managers
IF	Intermediate Frequency
IRS	Independent Rear Suspension
LEMA	Local Emergency Management Agency
LFE	Low-Frequency Earthquakes
LIGO	Laser Interferometer Gravitational-Wave Observatory
LIM	Linear Induction Motor
LO	Local Oscillator
LRFD	Load and Resistance Factor Design
LSM	Linear Synchronous Motor
MYD	Metallic Yielding Dampers
NEES	Network for Earthquake Engineering Simulation
NFPA	National Fire Protection Association
NSF	National Science Foundation
NSM	Negative-Stiffness-Mechanism
PML	Probable Maximum Loss
PZT	Piezoelectric Transducer
RGS	Remote Ground Sensors
SAMHSA	Substance Abuse and Mental Health Services Administration
SER	Smart Emergency Response System
SPSW	Steel Plate Shear Wall
SPT	Standard Penetration Test
SSE	Slow-Slip Events
SWLI	Scanning White Light Interferometry
TIEMS	The International Emergency Management Society
USGS	United States Geological Survey
VED	Viscoelastic Dampers
VSI	Vertical Scanning Interferometry

Preface

Earthquake engineering is a multi-phased process that ranges from the description of earthquake sources, to characterization of site effects and structural response, and to description of measures of seismic protection. Earthquake engineering is the science of the performance of buildings and structures when subjected to seismic loading. It also assists analyzing the interaction between civil infrastructure and the ground, including the consequences of earthquakes on structures. One of the most important aims of earthquake engineering is the proper design and construction of buildings in accordance with building codes, so as to minimize damage due to earthquakes. It is the earthquake engineer who ensures proper design of buildings so they will resist damage due to earthquakes, but at the same time not be unnecessarily expensive. The purpose of these technologies is to minimize the seismic effects on buildings and other infrastructure by the use of seismic control devices. When seismic waves start penetrating the base of the buildings from the ground level, the flow density of their energy reduces due to reflections and other reasons. However, the remaining waves possess significant potential for damage when they reach the superstructure. Vibration control devices assist in the reduction of the damaging effects, and enhance the seismic performance characteristics of the building. When the seismic waves penetrate a superstructure, these are dissipated by the use of dampers, or dispersed in a wide range of frequencies. Mass dampers are also employed to absorb the resonant wave frequencies of seismic waves, thus reducing the damaging effects. Seismic isolation techniques are sometimes used to partly suppress the flow of seismic energy into the superstructure by the insertion of pads into or beneath the load bearing elements in the base of the structure. Thus, the structure is protected from the damaging consequences of an earthquake by decoupling the structure from the shaking ground.

Content Coverage

Chapters One and *Two* focus on overview of earthquake engineering and types of earthquake, respectively. One of the most important aims of earthquake engineering is the proper design and construction of buildings in accordance with building codes, so as to minimize damage due to earthquakes.

Chapters Three and *Four* present coverage about causes of earthquake and earthquake measurements, separately. Earthquakes happen deep underground along tectonic plate boundaries. Tectonic plates are what make up the earth's crust, its outermost layer.

Chapters Five and *Six* focus on approaches of seismic base isolation system and vibration isolation, correspondingly. Base isolation is one of the most powerful tools of earthquake engineering pertaining to the passive structural vibration control technologies, but vibration isolation is the process of isolating an object, such as a piece of equipment, from the source of vibrations.

Chapters Seven and *Eight* present about earthquake-resistant structures designed to withstand earthquakes, and disaster management plans through which communities reduce vulnerability to hazards and cope with disasters, separately.

Chapters Nine and *Ten* are geotechnical engineering and seismic retrofitting, distinctly Geotechnical engineering uses principles of soil mechanics and rock mechanics to investigate subsurface conditions and materials, while seismic retrofitting is predominantly concerned with structural improvements to reduce the seismic hazard of using the structures.

CHAPTER 1

Earthquake Engineering: General Overview

Chapter Contents

- **Introduction**
- **Seismic Loading**
- **Seismic Performance**
- **Research for Earthquake Engineering**
- **Earthquake Simulation**
- **Seismic Design Requirements**
- **Earthquake-Resistant Construction**

Earthquake engineering is the science of the performance of buildings and structures when subjected to seismic loading. It also assists analyzing the interaction between civil infrastructure and the ground, including the consequences of earthquakes on structures. One of the most important aims of earthquake engineering is the proper design and construction of buildings in accordance with building codes, so as to minimize damage due to earthquakes.

INTRODUCTION

Earthquake engineering is an interdisciplinary branch of engineering that designs and analyzes structures, such as buildings and bridges, with earthquakes in mind. Its overall goal is to make such structures more resistant to earthquakes. An earthquake (or seismic) engineer aims to construct structures that will not be damaged in minor shaking and will avoid serious damage or collapse in a major earthquake. Earthquake engineering is the scientific field concerned with protecting society, the natural environment, and the man-made environment from earthquakes by limiting the seismic risk to socio-economically acceptable levels. Traditionally, it has been narrowly defined as the study of the behavior of structures and geo-structures subject to seismic loading; it is considered as a subset of structural engineering, geotechnical engineering, mechanical engineering, chemical engineering, applied physics, etc. However, the tremendous costs experienced in recent earthquakes have led to an expansion of its scope to encompass disciplines from the wider field of civil engineering, mechanical engineering and from the social sciences, especially sociology, political science, economics and finance.

Shake-table crash testing of a regular building model (left) and a base-isolated building model (right)at UCSD

The main objectives of earthquake engineering are:

- Foresee the potential consequences of strong earthquakes on urban areas and civil infrastructure.
- Design, construct and maintain structures to perform at earthquake exposure up to the expectations and in compliance with building codes.

A properly engineered structure does not necessarily have to be extremely strong or expensive. It has to be properly designed to withstand the seismic effects while sustaining an acceptable level of damage.

SEISMIC LOADING

Seismic loading means application of an earthquake-generated excitation on a structure (or geo-structure). It happens at contact surfaces of a structure either with the ground, with adjacent structures, or with gravity waves from tsunami. The loading that is expected at a given location on the Earth's surface is estimated by engineering seismology. It is related to the seismic hazard of the location.

Tokyo Skytree, equipped with a tuned mass damper, is the world's tallest tower and is the world's second tallest structure.

SEISMIC PERFORMANCE

Earthquake or seismic performance defines a structure's ability to sustain its main functions, such as its safety and serviceability, *at* and *after* a particular earthquake exposure. A structure is normally considered *safe* if it does not endanger the lives and well-being of those in or around it by partially or completely collapsing. A structure may be considered *serviceable* if it is able to fulfill its operational functions for which it was designed.

Basic concepts of the earthquake engineering, implemented in the major building codes, assume that a building should survive a rare, very severe earthquake by sustaining significant damage but without globally collapsing. On the other hand, it should remain operational for more frequent, but less severe seismic events.

Seismic Performance Assessment

Engineers need to know the quantified level of the actual or anticipated seismic performance associated with the direct damage to an individual building subject to a specified ground shaking. Such an assessment may be performed either experimentally or analytically.

Experimental Assessment

Experimental evaluations are expensive tests that are typically done by placing a (scaled) model of the structure on a shake-table that simulates the earth shaking and observing its behavior. Such kinds of experiments were first performed more than a century ago. Only recently has it become possible to perform 1:1 scale testing on full structures.

Due to the costly nature of such tests, they tend to be used mainly for understanding the seismic behavior of structures, validating models and verifying analysis methods. Thus, once properly validated, computational models and numerical procedures tend to carry the major burden for the seismic performance assessment of structures.

Analytical/Numerical Assessment

Seismic performance assessment or seismic structural analysis is a powerful tool of earthquake engineering which utilizes detailed modelling of the structure together with methods of structural analysis to gain a better understanding of seismic performance of building and non-building structures. The technique as a formal concept is a relatively recent development.

Snapshot from shake-table video of a 6-story non-ductile concrete building destructive testing

In general, seismic structural analysis is based on the methods of structural dynamics. For decades, the most prominent instrument of seismic analysis has been the earthquake response spectrum method which also contributed to the proposed building code's concept of today.

However, such methods are good only for linear elastic systems, being largely unable to model the structural behavior when damage (i.e., non-linearity) appears. Numerical *step-by-step integration* proved to be a more effective method of analysis for multi-degree-of-freedom structural systems with significant non-linearity under a transient process of ground motion excitation.

Basically, numerical analysis is conducted in order to evaluate the seismic performance of buildings. Performance evaluations are generally carried out by using nonlinear static pushover analysis or nonlinear time-history analysis. In such analyses, it is essential to achieve accurate non-linear modeling of structural components such as beams, columns, beam-column joints, shear walls etc. Thus, experimental results play an important role in determining the modeling parameters of individual components, especially those that are subject to significant non-linear deformations. The individual components are then assembled to create a full non-linear model of the structure. Thus created models are analyzed to evaluate the performance of buildings.

The capabilities of the structural analysis software are a major consideration in the above process as they restrict the possible component models, the analysis methods available and, most importantly, the numerical robustness. The latter becomes a major consideration for structures that venture into the non-linear range and approach global or local collapse as the numerical solution becomes increasingly unstable and thus difficult to reach. There are several commercially available Finite Element Analysis software's such as CSI-SAP2000 and CSI-PERFORM-3D and Scia Engineer-ECtools which can be used for the seismic performance evaluation of buildings. Moreover, there is research-based finite element analysis platforms such as OpenSees, RUAUMOKO and the older DRAIN-2D/3D, several of which are now open source.

RESEARCH FOR EARTHQUAKE ENGINEERING

Research for earthquake engineering means both field and analytical investigation or experimentation intended for discovery and scientific explanation of earthquake engineering related facts, revision of conventional concepts in the light of new findings, and practical application of the developed theories.

Shake-table testing of Friction Pendulum Bearings at EERC

The National Science Foundation (NSF) is the main United States government agency that supports fundamental research and education in all fields of earthquake engineering. In particular, it focuses on experimental, analytical and computational research on design and performance enhancement of structural systems.

E-Defense Shake Table

The Earthquake Engineering Research Institute (EERI) is a leader in dissemination of earthquake engineering research related information both in the U.S. and globally.

A definitive list of earthquake engineering research related shaking tables around the world may be found in Experimental Facilities for Earthquake Engineering Simulation Worldwide. The most prominent of them is now E-Defense Shake Table in Japan.

Major U.S. Research Programs

NSF also supports the George E. Brown, Jr. Network for Earthquake Engineering Simulation

The NSF Hazard Mitigation and Structural Engineering program (HMSE) supports research on new technologies for improving the behavior and response of structural systems subject to earthquake hazards; fundamental research on safety and reliability of constructed systems; innovative developments in analysis and model based simulation of structural behavior and response including soil-structure interaction; design concepts that improve structure performance and flexibility; and application of new control techniques for structural systems.

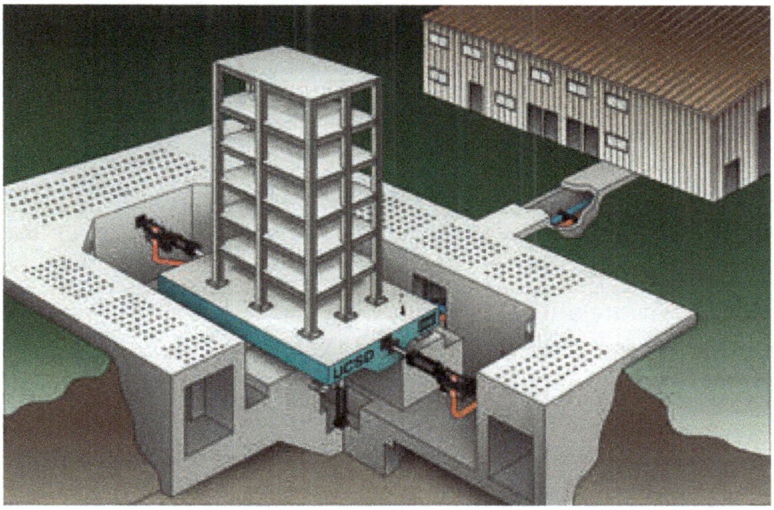

Large High Performance Outdoor Shake Table, UCSD, NEES network

(NEES) that advances knowledge discovery and innovation for earthquakes and tsunami loss reduction of the nation's civil infrastructure and new experimental simulation techniques and instrumentation.

The NEES network features 14 geographically-distributed, shared-use laboratories that support several types of experimental work:geotechnical centrifuge research, shake-table tests, large-scale structural testing, tsunami wave basin experiments, and field site research. Participating universities include: Cornell University; Lehigh University; Oregon State University; Rensselaer Polytechnic Institute; University at Buffalo, State University of New York; University of California, Berkeley; University of California, Davis; University of California, Los Angeles; University of California, San Diego; University of California, Santa Barbara; University of Illinois, Urbana-Champaign; University of Minnesota; University of Nevada, Reno; and the University of Texas, Austin.

NEES at Buffalo testing facility

The equipment sites (labs) and a central data repository are connected to the global earthquake engineering community via the NEEShub website. The NEES website is powered by HUBzero software developed at Purdue University for nanoHUB specifically to help the scientific community share resources and collaborate. The cyberinfrastructure, connected via Internet2, provides interactive simulation tools, a simulation tool development area, a curated central data repository, animated presentations, user support, telepresence, mechanism for uploading and sharing resources, and statistics about users and usage patterns.

This cyberinfrastructure allows researchers to: securely store, organize and share data within a standardized framework in a central location; remotely observe and participate in experiments through the use of synchronized real-time data and video; collaborate with colleagues to facilitate the planning, performance, analysis, and publication of research experiments; and conduct computational and hybrid simulations that may combine the results of multiple distributed experiments and link physical experiments with computer simulations to enable the investigation of overall system performance.

These resources jointly provide the means for collaboration and discovery to improve the seismic design and performance of civil and mechanical infrastructure systems.

EARTHQUAKE SIMULATION

The very first earthquake simulations were performed by statically applying some *horizontal inertia forces* based on scaled peak ground accelerations to a mathematical model of a building. With the further development of computational technologies, static approaches began to give way to dynamic ones.

Dynamic experiments on building and non-building structures may be physical, like shake-table testing, or virtual ones. In both cases, to verify a structure's expected seismic performance, some researchers prefer to deal with so called "real time-histories" though the last cannot be "real" for a hypothetical earthquake specified by either a building code or by some particular research requirements. Therefore, there is a strong incentive to engage an earthquake simulation which is the seismic input that possesses only essential features of a real event.

Sometimes earthquake simulation is understood as a re-creation of local effects of a strong earth shaking.

Structure Simulation

Theoretical or experimental evaluation of anticipated seismic performance mostly requires a structure simulation which is based on the concept of structural likeness or similarity. Similarity is some degree of analogy or resemblance between two or more objects. The notion of similarity rests either on exact or approximate repetitions of patterns in the compared items.

Concurrent experiments with two building models which are *kinematically equivalent* to a real prototype.

In general, a building model is said to have similarity with the real object if the two share *geometric similarity*, *kinematic similarity* and *dynamic similarity*. The most vivid and effective type of similarity is the *kinematic* one. *Kinematic similarity* exists when the paths and velocities of moving particles of a model and its prototype are similar.

The ultimate level of *kinematic similarity* is *kinematic equivalence* when, in the case of earthquake engineering, time-histories of each story lateral displacements of the model and its prototype would be the same.

Seismic Vibration Control

Seismic vibration control is a set of technical means aimed to mitigate seismic impacts in building and non-building structures. All seismic vibration control devices may be classified as *passive*, *active* or *hybrid* where:

- *passive control devices* have no feedback capability between them, structural elements and the ground;
- *active control devices* incorporate real-time recording instrumentation on the ground integrated with earthquake input processing equipment and actuators within the structure;
- *hybrid control devices* have combined features of active and passive control systems.

When ground seismic waves reach up and start to penetrate a base of a building, their energy flow density, due to reflections, reduces dramatically: usually, up to 90%. However, the remaining portions of the incident waves during a major earthquake still bear a huge devastating potential.

After the seismic waves enter a superstructure, there are a number of ways to control them in order to soothe their damaging effect and improve the building's seismic performance, for instance:

- to dissipate the wave energy inside a superstructure with properly engineered dampers;
- to disperse the wave energy between a wider range of frequencies;
- to absorb the resonant portions of the whole wave frequencies band with the help of so-called *mass dampers*.

Mausoleum of Cyrus, the oldest base-isolated structure in the world

Devices of the last kind, abbreviated correspondingly as TMD for the tuned (*passive*), as AMD for the *active*, and as HMD for the *hybrid mass dampers*, have been studied and installed in high-rise buildings, predominantly in Japan, for a quarter of a century.

However, there is quite another approach: partial suppression of the seismic energy flow into the superstructure known as seismic or base isolation.

For this, some pads are inserted into or under all major load-carrying elements in the base of the building which should substantially decouple a superstructure from its substructure resting on a shaking ground.

The first evidence of earthquake protection by using the principle of base isolation was discovered in Pasargadae, a city in ancient Persia, now Iran, and dates back to the 6th century BCE. Below, there are some samples of seismic vibration control technologies of today.

Dry-Stone Walls Control

People of Inca civilization were masters of the polished '*dry-stone walls*', called ashlar, where blocks of stone were cut to fit together tightly without any mortar. The Incas were among the best stonemasons the world has ever seen and many junctions in their masonry were so perfect that even blades of grass could not fit between the stones.

Dry-stone walls of Machu Picchu Temple of the Sun, Peru

Peru is a highly seismic land and for centuries the mortar-free construction proved to be apparently more earthquake-resistant than using mortar. The stones of the dry-stone walls built by the Incas could move slightly and resettle without the walls collapsing, a passive structural control technique employing both the principle of energy dissipation and that of suppressing resonant amplifications.

Tuned Mass Damper

Typically the tuned mass dampers are huge concrete blocks mounted in skyscrapers or other structures and moved in opposition to the resonance frequency oscillations of the structures by means of some sort of spring mechanism.

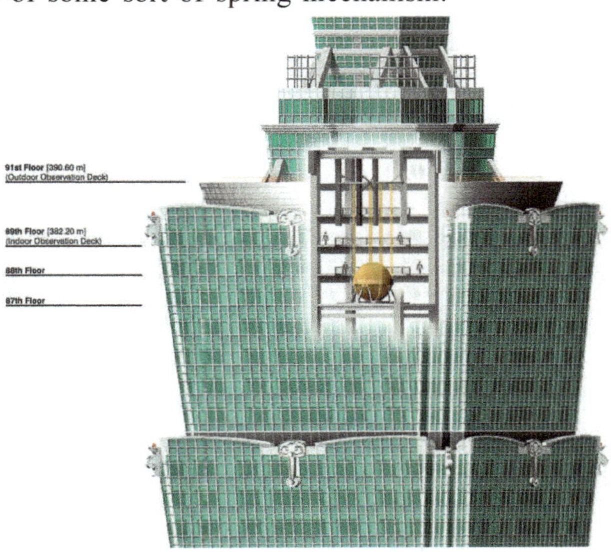

Tuned mass damper in Taipei 101, the world's third tallest skyscraper

Taipei 101 skyscraper needs to withstand typhoon winds and earthquake tremors common in its area of the Asia-Pacific. For this purpose, a steel pendulum weighing 660 metric tones that serves as a tuned mass damper was designed and installed atop the structure. Suspended from the 92nd to the 88th floor, the pendulum sways to decrease resonant amplifications of lateral displacements in the building caused by earthquakes and strong gusts.

Hysteretic Dampers

Hysteretic damper is intended to provide better and more reliable seismic performance than that of a conventional structure at the expense of the seismic input energy dissipation. There are five major groups of hysteretic dampers used for the purpose, namely:

- Fluid viscous dampers (FVDs)

Viscous Dampers have the benefit of being a supplemental damping system. They have an oval hysteretic loop and the damping is velocity dependent. While some minor maintenance is potentially required, viscous dampers generally do not need to be replaced after an earthquake. While more expensive than other damping technologies they can be used for both seismic and wind loads and are the most commonly used hysteric damper.

- Friction dampers (FDs)

Friction dampers tend to be available in two major types, linear and rotational and dissipate energy by heat. The damper operates on the principle of a coulomb damper. Depending on the design, friction dampers can experience stick-slip phenomenon and Cold welding. The main disadvantage being that friction surfaces can wear over time and for this reason they are not recommended for dissipating wind loads. When used in seismic applications wear is not a problem and there is no required maintenance. They have a rectangular hysteretic loop and as long as the building is sufficiently elastic they tend to settle back to their original positions after an earthquake.

- Metallic yielding dampers (MYDs)

Metallic yielding dampers, as the name implies, yield in order to absorb the earthquake's energy. This type of damper absorbs a large amount of energy however they must be replaced after an earthquake and may prevent the building from settling back to its original position.

- Viscoelastic dampers (VEDs)

Viscoelastic dampers are useful in that they can be used for both wind and seismic applications, they are usually limited to small displacements. There is some concern as to the reliability of the technology as some brands have been banned from use in buildings in the United States.

- Straddlingpendulum dampers (swing)

Base Isolation

Base Isolation seeks to prevent the kinetic energy of the earthquake from being transferred into elastic energy in the building. These technologies do so by isolating the structure from the ground, thus enabling them to move somewhat independently. The degree to which the energy is transferred into the structure and how the energy is dissipated will vary depending on the technology used.

- Lead Rubber Bearing

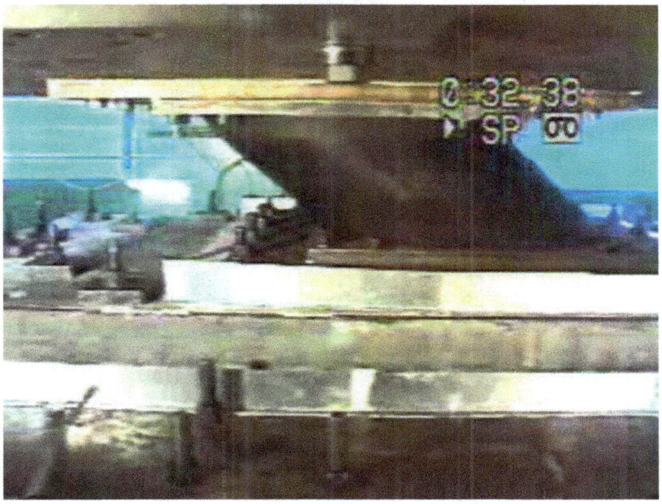

LRB being tested at the UCSD Caltrans-SRMD facility

Lead Rubber Bearing or LRB is a type of base isolation employing a heavy damping. It was invented by Bill Robinson, a New Zealander.

Heavy damping mechanism incorporated in vibration control technologies and, particularly, in base isolation devices, is often considered a valuable source of suppressing vibrations thus enhancing a building's seismic performance. However, for the rather pliant systems such as base isolated structures, with a relatively low bearing stiffness but with a high damping, the so-called "damping force" may turn out the main pushing force at a strong earthquake. The video shows a Lead Rubber Bearing being tested at the UCSD Caltrans-SRMD facility. The bearing is made of rubber with a lead core. It was a uniaxial test in which the bearing was also under a full structure load. Many buildings and bridges, both in New Zealand and elsewhere, are protected with lead dampers and lead and rubber bearings. Te Papa Tongarewa, the national museum of New Zealand, and the New Zealand Parliament Buildings have been fitted with the bearings. Both are in Wellington which sits on an active earthquake fault.

- Springs-with-damper base isolator

Springs-with-damper close-up

Springs-with-damper base isolator installed under a three-story town-house, Santa Monica, California is shown on the photo taken prior to the 1994 Northridge earthquake exposure. It is a base isolation device conceptually similar to *Lead Rubber Bearing*.

One of two three-story town-houses like this, which was well instrumented for recording of both vertical and horizontal accelerations on its floors and the ground, has survived a severe shaking during the Northridge earthquake and left valuable recorded information for further study.

- Simple roller bearing

Simple roller bearing is a base isolation device which is intended for protection of various building and non-building structures against potentially damaging lateral impacts of strong earthquakes.

This metallic bearing support may be adapted, with certain precautions, as a seismic isolator to skyscrapers and buildings on soft ground. Recently, it has been employed under the name of *Metallic Roller Bearing* for a housing complex (17 stories) in Tokyo, Japan.

- Friction pendulum bearing

Friction Pendulum Bearing (FPB) is another name of Friction Pendulum System (FPS). It is based on three pillars:

- articulated friction slider;
- spherical concave sliding surface;
- enclosing cylinder for lateral displacement restraint.

Snapshot with the link to video clip of a shake-table testing of FPB system supporting a rigid building model is presented at the right.

FPB shake-table testing

Seismic Design

Seismic design is based on authorized engineering procedures, principles and criteria meant to design or retrofit structures subject to earthquake exposure. Those criteria are only consistent with the contemporary state of the knowledge about earthquake engineering structures. Therefore, a building design which exactly follows seismic code regulations does not guarantee safety against collapse or serious damage.

The price of poor seismic design may be enormous. Nevertheless, seismic design has always been a trial and error process whether it was based on physical laws or on empirical knowledge of the structural performance of different shapes and materials.

San Francisco City Hall destroyed by 1906 earthquake and fire.

San Francisco after the 1906 earthquake and fire

To practice seismic design, seismic analysis or seismic evaluation of new and existing civil engineering projects, an engineer should, normally, pass examination on *Seismic Principles* which, in the State of California, include:

- Seismic Data and Seismic Design Criteria
- Seismic Characteristics of Engineered Systems
- Seismic Forces
- Seismic Analysis Procedures
- Seismic Detailing and Construction Quality Control

To build up complex structural systems, seismic design largely uses the same relatively small number of basic structural elements (to say nothing of vibration control devices) as any non-seismic design project.

Normally, according to building codes, structures are designed to "withstand" the largest earthquake of a certain probability that is likely to occur at their location. This means the loss of life should be minimized by preventing collapse of the buildings.

Seismic design is carried out by understanding the possible failure modes of a structure and providing the structure with appropriate strength, stiffness, ductility, and configuration to ensure those modes cannot occur.

SEISMIC DESIGN REQUIREMENTS

Seismic design requirements depend on the type of the structure, locality of the project and its authorities which stipulate applicable seismic design codes and criteria. For instance, California Department of Transportation's requirements called *The Seismic Design Criteria* (SDC) and aimed at the design of new bridges in California incorporate an innovative seismic performance-based approach.

The Metsamor Nuclear Power Plant was closed after the 1988 Armenian earthquake

The most significant feature in the SDC design philosophy is a shift from a *force-based assessment* of seismic demand to a *displacement-based assessment* of demand and capacity. Thus, the newly adopted displacement approach is based on comparing the *elastic displacement* demand to the *inelastic displacement* capacity of the primary structural components while ensuring a minimum level of inelastic capacity at all potential plastic hinge locations.

In addition to the designed structure itself, seismic design requirements may include a *ground stabilization* underneath the structure: sometimes, heavily shaken ground breaks up which leads to collapse of the structure sitting upon it. The following topics should be of primary concerns: liquefaction; dynamic lateral earth pressures on retaining walls; seismic slope stability; earthquake-induced settlement.

Nuclear facilities should not jeopardise their safety in case of earthquakes or other hostile external events. Therefore, their seismic design is based on criteria far more stringent than those applying to non-nuclear facilities. The Fukushima I nuclear accidents and damage to other nuclear facilities that followed the 2011 Tōhoku earthquake and tsunami have, however, drawn attention to ongoing concerns over Japanese nuclear seismic design standards and caused other many governments to re-evaluate their nuclear programs. Doubt has also been expressed over the seismic evaluation and design of certain other plants, including the Fessenheim Nuclear Power Plant in France.

Failure Modes

Failure mode is the manner by which an earthquake induced failure is observed. It, generally, describes the way the failure occurs. Though costly and time consuming, learning from each real earthquake failure remains a routine recipe for advancement in *seismic design* methods. Below, some typical modes of earthquake-generated failures are presented.

Typical damage to unreinforced masonry buildings at earthquakes

The lack of reinforcement coupled with poor mortar and inadequate roof-to-wall ties can result in substantial damage to an unreinforced masonry building. Severely cracked or leaning walls are some of the most common earthquake damage. Also hazardous is the damage that may occur between the walls and roof or floor diaphragms. Separation between the framing and the walls can jeopardize the vertical support of roof and floor systems.

Soft story collapse due to inadequate shear strength at ground level, Loma Prieta earthquake

Soft story effect. Absence of adequate shear walls on the ground level caused damage to this structure. A close examination of the image reveals that the rough board siding, once covered by a brick veneer, has been completely dismantled from the studwall. Only the rigidity of the floor above combined with the support on the two hidden sides by continuous walls, not penetrated with large doors as on the street sides, is preventing full collapse of the structure.

Effects of soil liquefaction during the 1964 Niigata earthquake

Soil liquefaction. In the cases where the soil consists of loose granular deposited materials with the tendency to develop excessive hydrostatic pore water pressure of sufficient magnitude and compact, liquefaction of those loose saturated deposits may result in non-uniform settlements and tilting of structures. This caused major damage to thousands of buildings in Niigata, Japan during the 1964 earthquake.

Car smashed by landslide rock, 2008 Sichuan earthquake

Landslide rock fall. A landslide is a geological phenomenon which includes a wide range of ground movement, including rock falls. Typically, the action of gravity is the primary driving force for a landslide to occur though in this case there was another contributing factor which affected the original slope stability: the landslide required an *earthquake trigger* before being released.

Effects of pounding against adjacent building, Loma Prieta

Pounding against adjacent building. This is a photograph of the collapsed five-story tower, St. Joseph's Seminary, Los Altos, California which resulted in one fatality. During Loma Prieta earthquake, the tower pounded against the independently vibrating adjacent building behind. A possibility of pounding depends on both buildings' lateral displacements which should be accurately estimated and accounted for.

Effects of completely shattered joints of concrete frame, Northridge

At Northridge earthquake, the Kaiser Permanente concrete frame office building had joints completely shattered, revealing inadequate confinement steel, which resulted in the second story collapse. In the transverse direction, composite end shear walls, consisting of two wythes of brick and a layer of shotcrete that carried the lateral load, peeled apart because of inadequate through-ties and failed.

7-story reinforced concrete buildings on steep slope collapse due to the following:

- Improper construction site on a foothill.
- Poor detailing of the reinforcement (lack of concrete confinement in the columns and at the beam-column joints, inadequate splice length).
- Seismically weak soft story at the first floor.
- Long cantilevers with heavy dead load.

shifting from foundation, Whittier

Sliding off foundations effect of a relatively rigid residential building structure during 1987 Whittier Narrows earthquake. The magnitude 5.9 earthquake pounded the Garvey West Apartment building in Monterey Park, California and shifted its superstructure about 10 inches to the east on its foundation.

Earthquake damage in Pichilemu.

If a superstructure is not mounted on a base isolation system, its shifting on the basement should be prevented.

Insufficient shear reinforcement let main rebars to buckle, Northridge

Reinforced concrete column burst at Northridge earthquake due to insufficient shear reinforcement mode which allows main reinforcement to buckle outwards. The deck unseated at the hinge and failed in shear. As a result, the La Cienega-Venice underpass section of the 10 Freeway collapsed.

Support-columns and upper deck failure, Loma Prieta earthquake

Loma Prieta earthquake: side view of reinforced concrete support-columns failure which triggered the upper deck collapse onto the lower deck of the two-level Cypress viaduct of Interstate Highway 880, Oakland, CA.

Failure of retaining wall due to ground movement, Loma Prieta

Retaining wall failure at Loma Prieta earthquake in Santa Cruz Mountains area: prominent northwest-trending extensional cracks up to 12 cm (4.7 in) wide in the concrete spillway to Austrian Dam, the north abutment.

Lateral spreading mode of ground failure, Loma Prieta

Ground shaking triggered soil liquefaction in a subsurface layer of sand, producing differential lateral and vertical movement in an overlying carapace of unliquified sand and silt. This mode of ground failure, termed lateral spreading, is a principal cause of liquefaction-related earthquake damage.

Beams and pier columns diagonal cracking, 2008 Sichuan earthquake

Severely damaged building of Agriculture Development Bank of China after 2008 Sichuan earthquake: most of the beams and pier columns are sheared. Large diagonal cracks in masonry and veneer are due to in-plane loads while abrupt settlement of the right end of the building should be attributed to a landfill which may be hazardous even without any earthquake.

Tsunami strikes Ao Nang,

Twofold tsunami impact: sea waves hydraulic pressure and inundation. Thus, the Indian Ocean earthquake of December 26, 2004, with the epicenter off the west coast of Sumatra, Indonesia, triggered a series of devastating tsunamis, killing more than 230,000 people in eleven countries by inundating surrounding coastal communities with huge waves up to 30 meters (100 feet) high.

EARTHQUAKE-RESISTANT CONSTRUCTION

Earthquake construction means implementation of seismic design to enable building and non-building structures to live through the anticipated earthquake exposure up to the expectations and in compliance with the applicable building codes.

Construction of Pearl River Tower X-bracing to resist lateral forces of earthquakes and winds

Design and construction are intimately related. To achieve a good workmanship, detailing of the members and their connections should be as simple as possible. As any construction in general, earthquake construction is a process that consists of the building, retrofitting or assembling of infrastructure given the construction materials available.

The destabilizing action of an earthquake on constructions may be *direct* (seismic motion of the ground) or *indirect* (earthquake-induced landslides, soil liquefaction and waves of tsunami).

A structure might have all the appearances of stability, yet offer nothing but danger when an earthquake occurs. The crucial fact is that, for safety, earthquake-resistant construction techniques are as important as quality control and using correct materials. *Earthquake contractor* should be registered in the state/province/country of the project location (depending on local regulations), bonded and insured.

To minimize possible losses, construction process should be organized with keeping in mind that earthquake may strike any time prior to the end of construction.

Each construction project requires a qualified team of professionals who understand the basic features of seismic performance of different structures as well as construction management.

Adobe Structures

Around thirty percent of the world's population lives or works in earth-made construction. Adobe type of mud bricks is one of the oldest and most widely used building materials. The use of adobe is very common in some of the world's most hazard-prone regions, traditionally across Latin America, Africa, Indian subcontinent and other parts of Asia, Middle East and Southern Europe.

Partially collapsed adobe building in Westmorland, California

Adobe buildings are considered very vulnerable at strong quakes. However, multiple ways of seismic strengthening of new and existing adobe buildings are available.

Key factors for the improved seismic performance of adobe construction are:

- Quality of construction.
- Compact, box-type layout.
- Seismic reinforcement.

Limestone and Sandstone Structures

Limestone is very common in architecture, especially in North America and Europe. Many landmarks across the world are made of limestone. Many medieval churches and castles in Europe are made of limestone and sandstone masonry. They are the long-lasting materials but their rather heavy weight is not beneficial for adequate seismic performance.

Base-isolated City and County Building, Salt Lake City, Utah

Application of modern technology to seismic retrofitting can enhance the survivability of unreinforced masonry structures. As an example, from 1973 to 1989, the Salt Lake City and County Building in Utah was exhaustively renovated and repaired with an emphasis on preserving historical accuracy in appearance. This was done in concert with a seismic upgrade that placed the weak sandstone structure on base isolation foundation to better protect it from earthquake damage.

Timber Frame Structures

Timber framing dates back thousands of years, and has been used in many parts of the world during various periods such as ancient Japan, Europe and medieval England in localities where timber was in good supply and building stone and the skills to work it were not.

Anne Hvide's House, Denmark (1560)

The use of timber framing in buildings provides their complete skeletal framing which offers some structural benefits as the timber frame, if properly engineered, lends itself to better *seismic survivability*.

Light-Frame Structures

Light-frame structures usually gain seismic resistance from rigid plywood shear walls and wood structural panel diaphragms. Special provisions for seismic load-resisting systems for all engineered wood structures requires consideration of diaphragm ratios, horizontal and vertical diaphragm shears, and connector/fastener values. In addition, collectors, or drag struts, to distribute shear along a diaphragm length are required.

A two-story wooden-frame for a residential building structure

Reinforced Masonry Structures

A construction system where steel reinforcement is embedded in the mortar joints of masonry or placed in holes and after filled with concrete or grout is called reinforced masonry.

The devastating 1933 Long Beach earthquake revealed that masonry construction should be improved immediately. Then, the California State Code made the reinforced masonry mandatory.

Reinforced hollow masonry wall

There are various practices and techniques to achieve reinforced masonry. The most common type is the reinforced hollow unit masonry. The effectiveness of both vertical and horizontal reinforcement strongly depends on the type and quality of the masonry, i.e. masonry units and mortar.

To achieve a ductile behavior of masonry, it is necessary that the shear strength of the wall is greater than the flexural strength.

Reinforced Concrete Structures

Reinforced concrete is concrete in which steel reinforcement bars (rebars) or fibers have been incorporated to strengthen a material that would otherwise be brittle. It can be used to produce beams, columns, floors or bridges.

Stressed Ribbon pedestrian bridge over the Rogue River, Grants Pass, Oregon

Prestressed concrete cable-stayed bridge over Yangtze river

Prestressed concrete is a kind of reinforced concrete used for overcoming concrete's natural weakness in tension. It can be applied to beams, floors or bridges with a longer span than is practical with ordinary reinforced concrete. Prestressing tendons (generally of high tensile steel cable or rods) are used to provide a clamping load which produces a compressive stress that offsets the tensile stress that the concrete compression member would, otherwise, experience due to a bending load.

To prevent catastrophic collapse in response earth shaking (in the interest of life safety), a traditional reinforced concrete frame should have ductile joints. Depending upon the methods used and the imposed seismic forces, such buildings may be immediately usable, require extensive repair, or may have to be demolished.

Prestressed Structures

Prestressed structure is the one whose overall integrity, stability and security depend, primarily, on a *prestressing*. *Prestressing* means the intentional creation of permanent stresses in a structure for the purpose of improving its performance under various service conditions.

Naturally pre-compressed exterior wall of Colosseum, Rome

There are the following basic types of prestressing:

- Pre-compression (mostly, with the own weight of a structure)
- Pretensioning with high-strength embedded tendons
- Post-tensioning with high-strength bonded or unbonded tendons

Today, the concept of prestressed structure is widely engaged in design of buildings, underground structures, TV towers, power stations, floating storage and offshore facilities, nuclear reactor vessels, and numerous kinds of bridge systems.

A beneficial idea of *prestressing* was, apparently, familiar to the ancient Rome architects; look, e.g., at the tall attic wall of Colosseum working as a stabilizing device for the wall piers beneath.

Steel Structures

Steel structures are considered mostly earthquake resistant but some failures have occurred.

A great number of welded steel moment-resisting frame buildings, which looked earthquake-proof, surprisingly experienced brittle behavior and were hazardously damaged in the 1994 Northridge earthquake. After that, the Federal Emergency Management Agency (FEMA) initiated development of repair techniques and new design approaches to minimize damage to steel moment frame buildings in future earthquakes.

Collapsed section of the San Francisco–Oakland Bay Bridge in response to Loma Prieta earthquake

For structural steel seismic design based on Load and Resistance Factor Design (LRFD) approach, it is very important to assess ability of a structure to develop and maintain its bearing resistance in the inelastic range. A measure of this ability is ductility, which may be observed in a *material itself*, in a *structural element*, or to a *whole structure*.

As a consequence of Northridge earthquake experience, the American Institute of Steel Construction has introduced AISC 358 "Pre-Qualified Connections for Special and intermediate Steel Moment Frames." The AISC Seismic Design Provisions require that all Steel Moment Resisting Frames employ either connections contained in AISC 358, or the use of connections that have been subjected to pre-qualifying cyclic testing.

Prediction of Earthquake Losses

Earthquake loss estimation is usually defined as a *Damage Ratio* (DR) which is a ratio of the earthquake damage repair cost to the total value of a building. *Probable Maximum Loss* (PML) is a common term used for earthquake loss estimation, but it lacks a precise definition. In 1999, ASTM E2026 'Standard Guide for the Estimation of Building Damageability in Earthquakes' was produced in order to standardize the nomenclature for seismic loss estimation, as well as establish guidelines as to the review process and qualifications of the reviewer.

Earthquake loss estimations are also referred to as *Seismic Risk Assessments*. The risk assessment process generally involves determining the probability of various ground motions coupled with the vulnerability or damage of the building under those ground motions. The results are defined as a percent of building replacement value.

REFERENCES

1. Bozorgnia, Yousef; Bertero, Vitelmo V. (2004). Earthquake Engineering: From Engineering Seismology to Performance-Based Engineering. CRC Press. ISBN 978-0-8493-1439-1.
2. Berg, Glen V. (1983). Seismic Design Codes and Procedures. EERI. ISBN 0-943198-25-9.
3. "Earthquake Protector: Shake Table Crash Testing". YouTube. Retrieved 2012-07-31.
4. http://www.civenv.unimelb.edu.au/ejse/Archives/Fulltext/2006/200608.pdf
5. Seismology Committee (1999). Recommended Lateral Force Requirements and Commentary. Structural Engineers Association of California.
6. neesit (2007-11-17). "Shaking Table Test on Conventional Wooden House (1)". YouTube. Retrieved 2012-07-31.
7. Omori, F. (1900). Seismic Experiments on the Fracturing and Overturning of Columns. Publ. Earthquake Invest. Comm. In Foreign Languages, N.4, Tokyo.
8. Chopra, Anil K. (1995). Dynamics of Structures. Prentice Hall. ISBN 0-13-855214-2.
9. Newmark, N.M.; Hall, W.J. (1982). Earthquake Spectra and Design. EERI. ISBN 0-943198-22-4.
10. Clough, Ray W.; Penzien, Joseph (1993). Dynamics of Structures. McGraw-Hill. ISBN 0-07-011394-7.
11. "Miki_house_test". YouTube. 2007-07-02. Retrieved 2012-07-31.
12. http://www.nea.fr/html/nsd/docs/2004/csni-r2004-10.pdf
13. "The NIED 'E-Defence' Laboratory in Miki City]". Retrieved 3 March 2008.
14. "CMMI - Funding - Hazard Mitigation and Structural Engineering - US National Science Foundation (NSF)". nsf.gov. Retrieved 2012-07-31.
15. "Network for Earthquake Engineering Simulation". Official web site. Retrieved September 21, 2011.
16. Lindeburg, Michael R.; Baradar, Majid (2001). Seismic Design of Building Structures. Professional Publications. ISBN 1-888577-52-5.
17. "Base isolation for earthquake engineering". YouTube. 2007-06-27. Retrieved

2012-07-31.

18. "Passive and active vibration isolation systems - Theory". Physics-animations.com. Retrieved 2012-07-31.
19. Chu, S.Y.; Soong, T.T.; Reinhorn, A.M. (2005). Active, Hybrid and Semi-Active Structural Control. John Wiley & Sons. ISBN 0-470-01352-4.
20. "Slide 2". Ffden-2.phys.uaf.edu. Retrieved 2012-07-31.
21. http://www.takenaka.co.jp/takenaka_e/quake_e/seishin/seishin.htm
22. "Live Event Q&As". Pbs.org. Retrieved 2013-07-28.
23. "Clark,Liesl; ''First Inhabitants'' PBS online, Nova; updated Nov. 2000". Pbs.org. Retrieved 2013-07-28.
24. Archived May 14, 2014, at the Wayback Machine.
25. "4. Building for earthquake resistance - Earthquakes - Te Ara Encyclopedia of New Zealand". Teara.govt.nz. 2009-03-02. Retrieved 2012-07-31.
26. neesit (2007-07-10). "LBRtest". YouTube. Retrieved 2012-07-31.
27. "Building Technology + Seismic Isolation System - OKUMURA CORPORATION" (in Japanese). Okumuragumi.co.jp. Retrieved 2012-07-31.
28. neesit (2007-04-19). "Hybrid Simulation of Base Isolated Structures". YouTube. Retrieved 2012-07-31.
29. Zayas, Victor A.; Low, Stanley S.; Mahin, Stephen A. (May 1990), "A Simple Pendulum Technique for Achieving Seismic Isolation", Earthquake Spectra, 6 (2): 317–333, doi:10.1193/1.1585573, ISSN 8755-2930
30. Housner, George W.; Jennings, Paul C. (1982). Earthquake Design Criteria. EERI. ISBN 1-888577-52-5.
31. "Earthquake-Resistant Construction". Nisee.berkeley.edu. Retrieved 2012-07-31.
32. http://www.pels.ca.gov/applicants/plan_civseism.pdf
33. Edited by Farzad Naeim (1989). Seismic Design Handbook. VNR. ISBN 0-442-26922-6.
34. Arnold, Christopher; Reitherman, Robert (1982). Building Configuration & Seismic Design. A Wiley-Interscience Publication. ISBN 0-471-86138-3.
35. "Template for External Caltrans Pages". Dot.ca.gov. Retrieved 2012-07-31.
36. "Strategy to Close Metsamor Plant Presented | Asbarez Armenian News". Asbarez.com. 1995-10-26. Retrieved 2012-07-31.
37. neesit. "Niigita Earthquake 1964 - YouTube". www.youtube.com. Retrieved 2012-07-31.
38. Robert W. Day (2007). Geotechnical Earthquake Engineering Handbook. McGraw Hill. ISBN 0-07-137782-4.
39. "Nuclear Power Plants and Earthquakes". World-nuclear.org. Retrieved 2013-07-28.
40. neesit. "Niigita Earthquake 1964". YouTube. Retrieved 2012-07-31.

41. "Soil Liquefaction with Dr. Ellen Rathje". YouTube. Retrieved 2013-07-28.
42. "Building Collapse". YouTube. Retrieved 2013-07-28.
43. "Tsunami disaster (Sri Lanka Resort)". YouTube. Retrieved 2013-07-28.
44. "YouTube". YouTube. Retrieved 2013-07-28.
45. Edited by Robert Lark (2007). Bridge Design, Construction and Maintenance. Thomas Telford. ISBN 0-7277-3593-4.
46. "Bad construction cited in quake zone - World news - Asia-Pacific - China earthquake | NBC News". MSNBC. Retrieved 2013-07-28.
47. "Earth Architecture - the Book, Synopsis". Retrieved 21 January 2010.
48. "simulacion terremoto peru-huaraz - casas de adobe - YouTube". Nz.youtube.com. 2006-06-24. Retrieved 2013-07-28.
49. Archived August 28, 2008, at the Wayback Machine.
50. "Shake table testing of adobe house (4A-S7 East) - YouTube". Nz.youtube.com. 2007-01-12. Retrieved 2013-07-28.
51. Timber Design & Construction Sourcebook=Gotz, Karl-Heinz et al. McGraw-Hall. 1989. ISBN 0-07-023851-0.
52. "SEESL". Nees.buffalo.edu. Retrieved 2013-07-28.
53. Rossen Rashkoff. "Reinforced Brick Masonry". Staff.city.ac.uk. Retrieved 2013-07-28.
54. Ekwueme, Chukwuma G.; Uzarski, Joe (2003). Seismic Design of Masonry Using the 1997 UBC. Concrete Masonry Association of California and Nevada.
55. Nilson, Arthur H. (1987). Design of Prestressed Concrete. John Wiley & Sons. ISBN 0-471-83072-0.
56. Nawy, Edward G. (1989). Prestressed Concrete. Prentice Hall. ISBN 0-13-698375-8.
57. Reitherman, Robert (2012). Earthquakes and Engineers: An International History. Reston, VA: ASCE Press. pp. 394–395. ISBN 9780784410714.
58. "SAC Steel Project: Welcome". Sacsteel.org. Retrieved 2013-07-28.
59. Seismic Design Manual. Chicago: American Institute of Steel Construction. 2006. pp. 6.1–30. ISBN 1-56424-056-8.
60. EERI Endowment Subcommittee (May 2000). Financial Management of Earthquake Risk. EERI Publication. ISBN 0-943198-21-6.
61. Eugene Trahern (1999). "Loss Estimation".
62. Craig Taylor; Erik VanMarcke, eds. (2002). Acceptable Risk Processes: Lifeline and Natural Hazards. Reston, VA: ASCE, TCLEE. ISBN 9780784406236.

CHAPTER 2

Types of Earthquake

Chapter Contents

- **Introduction**
- **Foreshock**
- **Aftershock**
- **Blind Thrust Earthquake**
- **Doublet Earthquake**
- **Interplate Earthquake**
- **Intraplate Earthquake**
- **Megathrust Earthquake**
- **Remotely Triggered Earthquakes**
- **Slow Earthquake**
- **Submarine Earthquake**
- **Supershear Earthquake**
- **Tsunami Earthquake**

Earthquakes today and most of their observed properties are explained in terms of physical theory. Most earthquakes occur near plate margins, geologists conclude that global geologic, or tectonic forces that produce mountains, rift valleys, midoceanic ridges, and ocean trenches are also the underlying cause of great earthquakes. These global forces, which at present are still not fully understood are consequences of different temperatures within the earth. Different losses of heat by radiation into space and gain of heat from decay of radioactive elements within rock formations.

INTRODUCTION

Earthquakes can come in three main forms, depending on the plate movements that occur beneath the earth's surface. They could occur on a Convergent Boundary, Divergent Boundary or a Transform Fault.

The cause or source zone is sometimes used as a way to categorize earthquakes. When earthquakes are categorized by cause, types include tectonic earthquakes, volcanic earthquakes and explosive earthquakes. Explosive earthquakes are often man-made, and the two terms may be used interchangeably. When source zone is used to categorize earthquakes, the primary earthquake types are shallow-fault, deep zone and crustal earthquakes.

FORESHOCK

A foreshock is an earthquake that occurs before a larger seismic event (the *mainshock*) and is related to it in both time and space. The designation of an earthquake as *foreshock*, *mainshock* or aftershock is only possible after the full sequence of events has happened.

Occurrence

Foreshock activity has been detected for about 40% of all moderate to large earthquakes, and about 70% for events of M>7.0. They occur from a matter of minutes to days or even longer before the main shock, for example the 2002 Sumatra earthquake is regarded as a foreshock of the 2004 Indian Ocean earthquake with a delay of more than two years between the two events. Some great earthquakes (M>8.0) show no foreshock activity at all, such as the M8.6 1950 India - China earthquake. The increase in foreshock activity is difficult to quantify for individual earthquakes but becomes apparent when combining the results of many different events. From such combined observations, the increase before the mainshock is observed to be of inverse power law type. This may either indicate that foreshocks cause stress changes resulting in the mainshock or that the increase is related to a general increase in stress in the region.

Mechanics

The observation of foreshocks associated with many earthquakes suggests that they are part of a preparation process prior to nucleation. In one model of earthquake rupture, the process forms as a cascade, starting with a very small event that triggers a larger one, continuing until the main shock rupture is triggered. However, analysis of some foreshocks has shown that they tend to relieve stress around the fault. In this view, foreshocks and aftershocks are part of the same process. This is supported by an observed relationship between the rate of foreshocks and the rate of aftershocks for an event.

Earthquake Prediction

An increase in seismic activity in an area has been used as a method of predicting earthquakes, most notably in the case of the 1975 Haicheng earthquake in China, where an evacuation was triggered by an increase in activity. However, most earthquakes lack obvious foreshock patterns and this method has not proven useful, as most small earthquakes are not foreshocks, leading to probable false alarms. Earthquakes along oceanic transform faults do show repeatable foreshock behavior, allowing the prediction of both the location and timing of such earthquakes.

Examples of Earthquakes with Foreshock Events

- The latest example of these types of earthquake was the 2016 Kumamoto earthquakes.
- The strongest earthquake of this type is the 1960 Valdivia earthquake which had a magnitude of 9.5 M_W.

Date (Foreshock)	Magnitude (Foreshock)	Flag and Country	Region	Date	Depth	Magnitude	Intensity	Name	Deceased	Tsunami
May 21, 1960 (1 day)	7.9 M_W	Chile	Araucanía Region	May 22, 1960	35 km	9.5 M_W	XII Mercalli	1960 Valdivia earthquake	1,655	✓
November 2, 2002 (2 years)	7.3 M_W	Indonesia	Sumatra	December 26, 2004	30 km	9.1 M_W	✕	2004 Indian Ocean earthquake and tsunami	230,000	✓
October 20, 2006 (299 days)	6.4 M_W	Peru	Ica Region	August 15, 2007	35 km	8.0 M_W	VIII Mercalli	2007 Peru earthquake	596	✓
January 23, 2007 (3 months)	5.2 M_L	Chile	Aysén Region	April 21, 2007	6 km	6.2 M_W	VII Mercalli	2007 Aysén Fjord earthquake	10	✓
March 9, 2011 (2 days)	7.3 M_W	Japan	Miyagi Prefecture	March 11, 2011	30 km	9.0 M_W	IX Mercalli and 7 Shindo	2011 Tōhoku earthquake and tsunami	15,891	✓
March 16, 2014 (15 days)	6.7 M_W	Chile	Tarapacá Region	April 1, 2014	20.1 km	8.2 M_W	VIII Mercalli	2014 Iquique earthquake	7	✓
April 14, 2016 (2 days)	6.2 M_W	Japan	Kumamoto Prefecture	April 16, 2016	11 km	7.0 M_W	IX Mercalli	2016 Kumamoto earthquakes	41	X

Note: dates are in local time

AFTERSHOCK

An aftershock is a smaller earthquake that occurs after a previous large earthquake, in the same area of the main shock. If an aftershock is larger than the main shock, the aftershock is redesignated as the main shock and the original main shock is redesignated as a foreshock. Aftershocks are formed as the crust around the displaced fault plane adjusts to the effects of the main shock.

Distribution of Aftershocks

Most aftershocks are located over the full area of fault rupture and either occur along the fault plane itself or along other faults within the volume affected by the strain associated with the main shock. Typically, aftershocks are found up to a distance equal to the rupture length away from the fault plane.

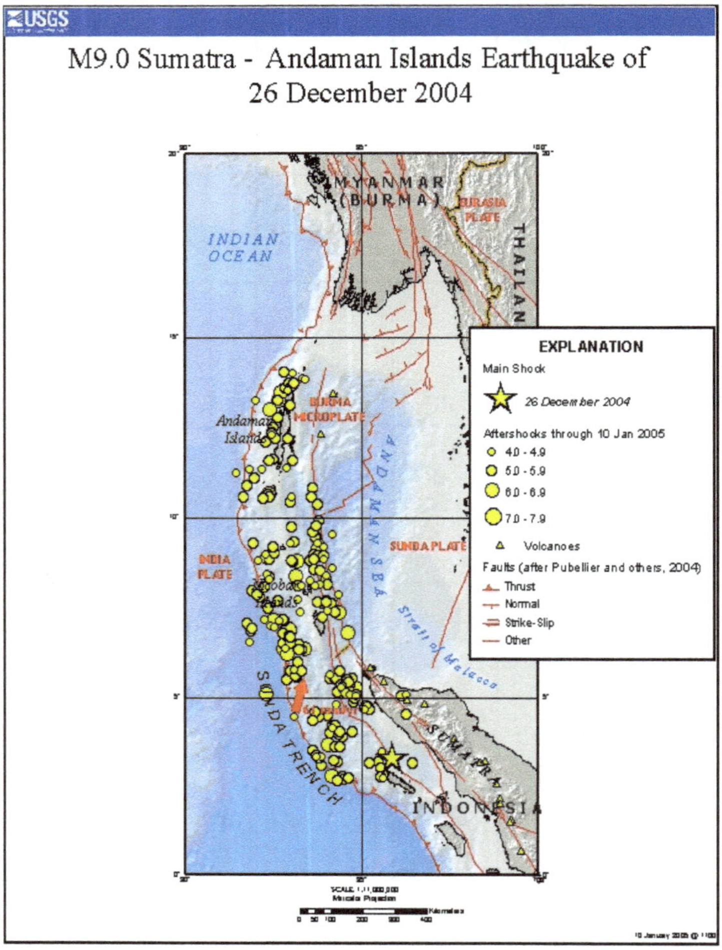

Aftershocks of the 2004 Indian Ocean earthquake, with data up to January 10, 2005.

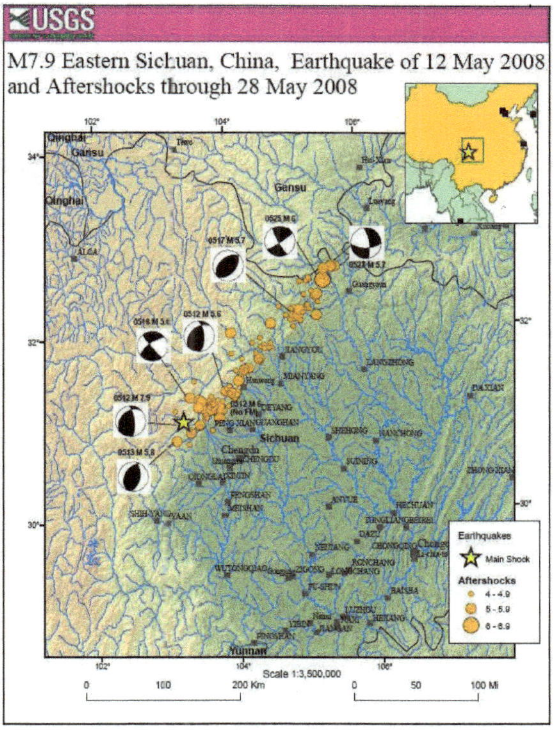

Distribution of aftershocks associated with the May 12th 2008 Sichuan earthquake through to 28th May

The pattern of aftershocks helps confirm the size of area that slipped during the main shock. In the case of the 2004 Indian Ocean earthquake and the 2008 Sichuan earthquake the aftershock distribution shows in both cases that the epicenter (where the rupture initiated) lies to one end of the final area of slip, implying strongly asymmetric rupture propagation.

Aftershock Size and Frequency with Time

Aftershocks rates and magnitudes follow several well-established empirical laws.

Omori's Law

The frequency of aftershocks decreases roughly with the reciprocal of time after the main shock. This empirical relation was first described by Fusakichi Omori in 1894 and is known as Omori's law. It is expressed as

$$n(t) = \frac{k}{(c+t)}$$

where k and c are constants, which vary between earthquake sequences. A modified version of Omori's law, now commonly used, was proposed by Utsu in 1961.

$$n(t) = \frac{k}{(c+t)^p}$$

where p is a third constant which modifies the decay rate and typically falls in the range 0.7–1.5.

According to these equations, the rate of aftershocks decreases quickly with time. The rate of aftershocks is proportional to the inverse of time since the mainshock and this relationship can be used to estimate the probability of future aftershock occurrence. Thus whatever the probability of an aftershock are on the first day, the second day will have 1/2 the probability of the first day and the tenth day will have approximately 1/10 the probability of the first day (when p is equal to 1). These patterns describe only the statistical behavior of aftershocks; the actual times, numbers and locations of the aftershocks are stochastic, while tending to follow these patterns. As this is an empirical law, values of the parameters are obtained by fitting to data after a mainshock has occurred, and they imply no specific physical mechanism in any given case.

Båth's Law

The other main law describing aftershocks is known as Båth's Law and this states that the difference in magnitude between a main shock and its largest aftershock is approximately constant, independent of the main shock magnitude, typically 1.1–1.2 on the Moment magnitude scale.

Gutenberg–Richter law

Aftershock sequences also typically follow the Gutenberg–Richter law of size scaling, which refers to the relationship between the magnitude and total number of earthquakes in a region in a given time period.

$$N = 10^{a-bM}$$

Where:

- N is the number of events greater or equal to M
- M is magnitude
- a and b are constants

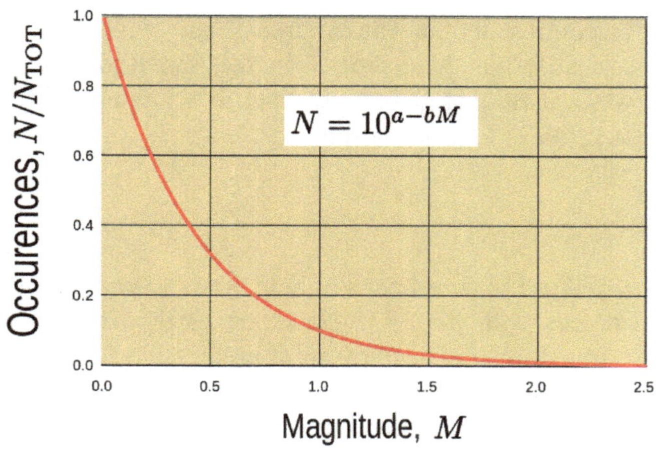

Gutenberg–Richter law for $b = 1$

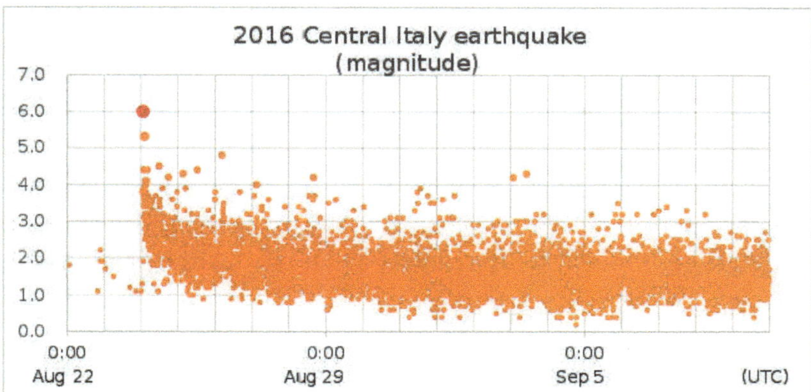

Magnitude of the Central Italy earthquake of August 2016 (red dot) and aftershocks (which continued to occur after the period shown here)

In summary, there are more small aftershocks and fewer large aftershocks.

Effect of Aftershocks

Aftershocks are dangerous because they are usually unpredictable, can be of a large magnitude, and can collapse buildings that are damaged from the main shock. Bigger earthquakes have more and larger aftershocks and the sequences can last for years or even longer especially when a large event occurs in a seismically quiet area; see, for example, the New Madrid Seismic Zone, where events still follow Omori's law from the main shocks of 1811–1812. An aftershock sequence is deemed to have ended when the rate of seismicity drops back to a background level; i.e., no further decay in the number of events with time can be detected.

Land movement around the New Madrid is reported to be no more than 0.2 mm (0.0079 in) a year, in contrast to the San Andreas Fault which averages up to 37 mm (1.5 in) a year across California. Aftershocks on the San Andreas are now believed to top out at 10 years while earthquakes in New Madrid are considered aftershocks nearly 200 years after the 1812 New Madrid earthquake.

Foreshocks

Some scientists have tried to use foreshocks to help predict upcoming earthquakes, having one of their few successes with the 1975 Haicheng earthquake in China. On the East Pacific Rise however, transform faults show quite predictable foreshock behaviour before the main seismic event. Reviews of data of past events and their foreshocks showed that they have a low number of aftershocks and high foreshock rates compared to continental strike-slip faults.

Modeling

Seismologists use tools such as the Epidemic-Type Aftershock Sequence model (ETAS) to study cascading aftershocks.

Psychology

Following a large earthquake and aftershocks, many people have reported feeling "phantom earthquakes" when in fact no earthquake was taking place. This condition, known as "earthquake sickness" is thought to be related to motion sickness, and usually goes away as seismic activity tails off.

BLIND THRUST EARTHQUAKE

A blind thrust earthquake occurs along a thrust fault that does not show signs on the Earth's surface, hence the designation "blind". Such faults, being invisible at the surface, have not been mapped by standard surface geological mapping. Sometimes they are discovered as a by-product of oil exploration seismology; in other cases their existence is not suspected.

Although such earthquakes are not amongst the most energetic, they are sometimes the most destructive, as conditions combine to form an urban earthquake which greatly affects urban seismic risk.

Blind Thrust Faults

Blind thrust faults generally exist near tectonic plate margins, in the broad disturbance zone. They form when a section of the Earth's crust is under high compressive stresses, due to plate margin collision, or the general geometry of how the plates are sliding past each other.

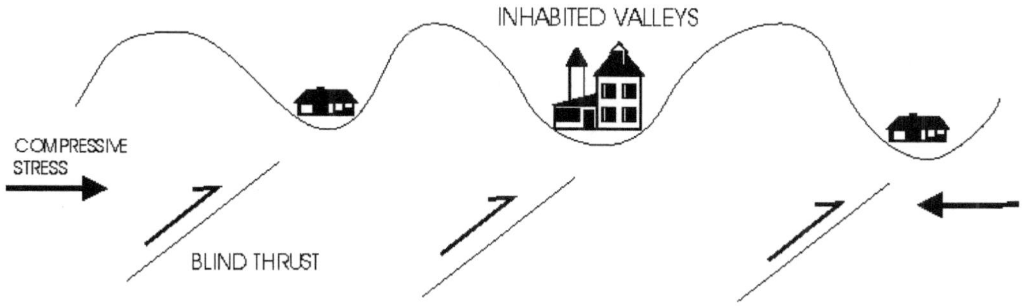

Diagram of blind-thrust faulting

As shown in the diagram, a weak plate under compression generally forms thrusting sheets, or overlapping sliding sections. This can form a hill and valley landform, with the hills being the strong sections, and the valleys being the highly disturbed thrust faulted and folded sections. After a long period of erosion the visible landscape may be flattened, with material eroded from the hills filling up the valleys and hiding the underlying hill-and-valley geology. The valley rock is very weak and usually highly weathered, presenting deep, fertile soil; naturally, this is the area that becomes populated. Reflection seismology profiles show the disturbed rock that hides a blind thrust fault.

If the region is under active compression these faults are constantly rupturing, but any given valley might only experience a large earthquake every few hundred years. Although usually of magnitude 6 to 7 compared to the largest magnitude 9 earthquakes of recent times, such a temblor is especially destructive because the seismic waves are highly directed, and the soft basin soil of the valley can amplify the ground motions tenfold or more.

It is said that blind thrust earthquakes contribute more to urban seismic risk than the 'big ones' of magnitude 8 or more.

Examples of Occurrence

Some Known Faults

- Los Angeles, California, USA, has many earthquakes and is well-studied. In addition to surface faults, a number of blind-thrust faults have been found under the basin and metropolitan area. A NASA study which combined satellite radar images and Global Positioning System (GPS) observations found that "tectonic squeezing across Los Angeles" "will likely produce earthquakes on either the blind Elysian Park or Puente Hills thrust fault systems".
- Bajo Segura Fault Zone, Spain
- Fukaya Fault System, Japan (near Tokyo)
- Uemachi Fault System, Osaka Basin, Japan

DOUBLET EARTHQUAKE

Seismologists sometimes refer to a pair of similarly sized earthquake shocks that occur relatively closely spaced in time and location as an earthquake "doublet." This is distinct from the normal pattern of earthquake aftershocks. Aftershocks gradually diminish in magnitude and generally come from the same origin as a mainshock, whereas doublet earthquakes originate from a place other than the original earthquake. The first earthquake can be some considerable distance and time away from the second. The magnitude of the second quake may be slightly larger than the first. This type of earthquake occurs once or twice every year but are far rarer than more typical earthquakes. In earthquake prone regions officials do not plan on doublet earthquakes occurring since they are a rare occurrence. However, when they do occur they have a high disaster potential. A relatively recent doublet earthquake occurred late in 2006 and early in 2007 in the Kuril Islands of Japan, which had not had a large scale earthquake since 1915. The first earthquake took place on November 15 and was of magnitude 8.3. Shortly after this seismic activity began in the Pacific plate where the second earthquake would take place on January 13, measuring a magnitude of 8.1. There was only 1 reported injury from the first earthquake and no reported injuries from the second earthquake since they both originated in the ocean. However, each earthquake caused a tsunami and the earthquake on November 15 created a tsunami that reached the coast of California causing $500,000- $1,000,000 in damages. There may also be examples of triplet earthquakes such as the 2010 Mindanao earthquakes.

INTERPLATE EARTHQUAKE

An interplate earthquake is an earthquake that occurs at the boundary between two tectonic plates. Earthquakes of this type account for more than 90 percent of the total seismic energy released around the world. If one plate is trying to move past the other, they will be locked until sufficient stress builds up to cause the plates to slip relative to each other. The slipping process creates an earthquake with land deformations and resulting seismic waves which travel through the Earth and along the Earth's surface. Relative plate motion can be lateral as along a transform fault boundary or vertical if along a convergent subduction boundary or a rift at a divergent boundary. At a subduction boundary the motion is due to one plate slipping beneath the other plate resulting in an interplate thrust or megathrust earthquake, which are the most powerful earthquakes.

Some areas of the world that are particularly prone to such events include the west coast of North America (especially California and Alaska), the northeastern Mediterranean region (Greece, Italy, and Turkey in particular), Iran, New Zealand, Indonesia, India, Japan, and parts of China.

Interplate earthquakes differ from intraplate earthquake in the intensity of stress drop which occurs after the quake. Intraplate earthquake have, on average, more stress drop than that of the interplate earthquake.Interplate earthquakes also differ fundamentally from intraplate earthquakes in the way stress is released and recovered. An interplate earthquake results in an immediate stress drop along the fault. Following this is a period of postseismic stress restoration. This restoration occurs quickly within the first few decades following the rupture and is due to tectonic loading and viscous relaxation in the lower crust. This results in a transfer of stress to the upper crust. Later on, a period of steady stress increase occurs due to tectonic loading.

INTRAPLATE EARTHQUAKE

An intraplate earthquake occurs in the interior of a tectonic plate, whereas an interplate earthquake is one that occurs at a plate boundary.

Intraplate earthquakes are relatively rare. Interplate earthquakes, which occur at plate boundaries, are more common. Nonetheless, very large intraplate earthquakes can inflict heavy damage, particularly because such areas are not accustomed to earthquakes and buildings are usually not seismically retrofitted. Examples of damaging intraplate earthquakes are the devastating Gujarat earthquake in 2001, the 2012 Indian Ocean earthquakes, the 1811-1812 earthquakes in New Madrid, Missouri, and the 1886 earthquake in Charleston, South Carolina.

Fault zones within Tectonic Plates

The surface of the Earth is made up of seven primary and eight secondary tectonic plates, plus dozens of tertiary microplates. The large plates move very slowly, owing to convection currents within the mantle below the crust. Because they do not all move in the same direction, plates often directly collide or move laterally along each other, a tectonic environment that makes earthquakes frequent. Relatively few earthquakes occur in intraplate environments; most occur on faults near plate margins. By definition, intraplate earthquakes do not occur near plate boundaries, but along faults in the normally stable interior of plates. These earthquakes often occur at the location of ancient failed rifts, because such old structures may present a weakness in the crust where it can easily slip to accommodate regional tectonic strain.

Compared to earthquakes near plate boundaries, intraplate earthquakes are not well understood, and the hazards associated with them may be difficult to quantify.

Historic Examples

Historic examples of intraplate earthquakes include those in Mineral, Virginia in 2011 (estimated magnitude 5.8), New Madrid in 1811 and 1812 (estimated magnitude as high as 8.1), the Boston (Cape Ann) earthquake of 1755 (estimated magnitude 6.0 to 6.3), earthquakes felt in New York City in 1737 and 1884 (both quakes estimated at about 5.5 magnitude), and the Charleston earthquake in South Carolina in 1886 (estimated magnitude 6.5 to 7.3). The Charleston quake was particularly surprising because, unlike Boston and New York, the area had almost no history of even minor earthquakes.

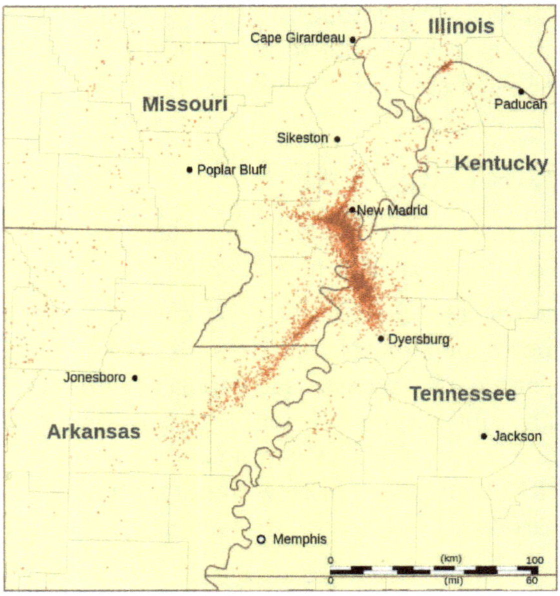

Distribution of seismicity associated with the New Madrid Seismic Zone (since 1974). This zone of intense earthquake activity is located deep in the interior of the North American plate.

In 2001, a large intraplate earthquake devastated the region of Gujarat, India. The earthquake occurred far from any plate boundaries, which meant the region above the epicenter was unprepared for earthquakes. In particular, the Kutch district suffered tremendous damage, where the death toll was over 12,000 and the total death toll was higher than 20,000.

Causes

Many cities live with the seismic risk of a rare, large intraplate earthquake. The cause of these earthquakes is often uncertain. In many cases, the causative fault is deeply buried, and sometimes cannot even be found. Under these circumstances it is difficult to calculate the exact seismic hazard for a given city, especially if there was only one earthquake in historical times. Some progress is being made in understanding the fault mechanics driving these earthquakes.

Prediction

Scientists continue to search for the causes of these earthquakes, and especially for some indication of how often they recur. The best success has come with detailed micro-seismic monitoring, involving dense arrays of seismometers. In this manner, very small earthquakes associated with a causative fault can be located with great accuracy, and in most cases these line up in patterns consistent with faulting. Cryoseisms can sometimes be mistaken for intraplate earthquakes.

MEGATHRUST EARTHQUAKE

Megathrust earthquakes occur at subduction zones at destructive convergent plate boundaries, where one tectonic plate is forced underneath another. These interplate earthquakes are the planet's most powerful, with moment magnitudes (M_w) that can exceed 9.0. Since 1900, all earthquakes of magnitude 9.0 or greater have been megathrust earthquakes. No other type of known terrestrial source of tectonic activity has produced earthquakes of this scale.

Terminology

During the rupture, one side of the fault is pushed upwards relative to the other, and it is this type of movement that is known as thrust. They are a type of dip-slip fault. A thrust fault is a reverse fault with a dip of 45° or less. Oblique-slip faults have significant components of different slip styles. The term *megathrust* does not have a widely accepted rigorous definition, but is used to refer to an extremely large thrust fault, typically formed at the plate interface along a subduction zone such as the Sunda megathrust. It is mostly American terminology.

Areas

The major subduction zone is associated with the Pacific and Indian Oceans and is responsible for the volcanic activity associated with the Pacific Ring of Fire. Since these earthquakes deform the ocean floor, they often generate a significant series of tsunami waves. They are known to produce intense shaking for periods of time that can last for up to a few minutes.

In Japan, the Nankai megathrust under the Nankai Trough is responsible for Nankai megathrust earthquakes and associated tsunamis.

A study reported in 2016 found that the largest megathrust quakes are associated with downgoing slabs with the shallowest dip, so-called "flat slab subduction".

Examples

Examples of megathrust earthquakes are listed in the following table.

Event	Estimated Magnitude (M_w)	Tectonic Plates Involved	Other Details/Notes
365 Crete earthquake	8.0+	African Plate subducting beneath the Aegean Sea Plate	• The quake generated a large tsunami in the eastern Mediterranean Sea and caused significant vertical displacement in the island of Crete.

869 Sanriku earthquake	8.6–9.0	Pacific Plate subducting beneath the Okhotsk Plate	• Slip length: 200 km over (125 mi over) • Slip width: 85 km over (53 mi over)
1575 Valdivia earthquake	8.5	Nazca Plate subducting beneath the South American Plate	
1700 Cascadia earthquake	8.7–9.2	Juan de Fuca Plate subducting beneath the North American Plate	• Slip length: 1000 km (625 mi) • Slip motion: 20 m (60 ft)
1707 Hōei earthquake	8.6–9.3	Philippine Sea Plate subducting beneath the Eurasian Plate	• Duration: approximately 10 minutes • Slip length: maybe 600 and 700 km (370 and 435 mi)
1730 Valparaíso earthquake	8.7-9.0	Nazca Plate subducting beneath the South American Plate	
1737 Kamchatka earthquake	8.3–9.0	Pacific Plate subducting beneath the Okhotsk Plate	• Duration: 15 minutes • Depth: 40 km • Slip length: maybe 700 km over (435 mi over)
1755 Lisbon earthquake	8.5–9.0	Hypothesized to be part of a young subduction zone but origin still debated. Related to the Azores–Gibraltar Transform Fault	• Destroyed Lisbon and was followed by a 20 metre high tsunami and many fires. • It caused 10,000-100,000 deaths. • The grave of Nuno Álvares Pereira was destroyed, along with the historical records of the voyages of Vasco da Gama and Christopher Columbus
1868 Arica earthquake	8.5–9.0	Nazca Plate subducting beneath the South American Plate	• Slip length: 600 km (370 mi)

1877 Iquique earthquake	8.5–9.0?	Nazca Plate subducting beneath the South American Plate	• Slip length: 420 and 450 km (230 and 245 mi)
1906 Ecuador–Colombia earthquake	8.8	Nazca Plate subducting beneath the South American Plate	
1946 Nankaidō earthquake	8.1	Philippine Sea Plate subducting beneath the Eurasian Plate	• Slip length: maybe 300 km (190 mi)
1952 Kamchatka earthquake	9.0	Pacific Plate subducting beneath the Okhotsk Plate	• Depth: 30 km • Slip length: maybe 600 km over (370 mi over)
1957 Andreanof Islands earthquake	8.6	Pacific Plate subducting beneath the North American Plate	• Depth: 33 km • Slip length: maybe 700 km over (435 mi over)
1960 Great Chilean earthquake	9.5	Nazca Plate subducting beneath the South American Plate	• Duration: 5–6 minutes • Depth: 33 km • Slip length: 850 and 1000 km (530 and 625 mi) • Slip width: 200 km (125 mi) • Slip motion: 20 m (60 ft)
1964 Alaska earthquake ("Good Friday" earthquake)	9.2	Pacific Plate subducting beneath the North American Plate	• Duration: 4–5 minutes • Depth: 25 km • Slip length: 800 and 850 km (500 and 530 mi) • Slip width: 250 km (155 mi) • Slip motion: 23 m (70 ft)

2001 southern Peru earthquake	8.4	Nazca Plate subducting beneath the South American Plate	• Depth: 33 km • Slip length: 200 km (125 mi) • Slip width: 110 km (70 mi)
2004 Sumatra-Andaman earthquake ("Indian Ocean earthquake")	9.1–9.3	India Plate subducting beneath the Burma Plate	• The total vertical displacement measured by sonar survey is about 40 m in the vicinity of the epicenter and occurred as two separate movements which created two large, steep, almost vertical cliffs, one above the other. • Duration: 8–10 minutes • Depth: 30 km • Slip length: 1000 and 1300 km (625 and 810 mi) • Slip width: 180 km (110 mi) • Slip motion: 33 m (110 ft)
2010 Chile earthquake	8.8	Nazca Plate subducting beneath the South American Plate	• Depth: 35 km • Slip length: 500 km (310 mi) • Slip width: 200 km (125 mi)
2011 Tōhoku earthquake and tsunami	9.1	Pacific Plate subducting beneath the Okhotsk Plate	• Duration: 6 minutes • Depth: 29 km • Slip length: 500 km (310 mi) • Slip width: 200 km (125 mi) • Slip motion: 20 m (60 ft)

2014 Iquique earthquake	8.2	Nazca Plate subducting beneath the South American Plate	• Depth: 20.1 km • Slip length: 170 km (105 mi) • Slip width: 70 km (45 mi)
2015 Illapel earthquake	8.3	Nazca Plate subducting beneath the South American Plate	• Depth: 25 km • Slip length: 260 km (160 mi) • Slip width: 80 km (50 mi)

REMOTELY TRIGGERED EARTHQUAKES

Remotely triggered earthquakes are a result of the effects of large earthquakes at considerable distance, outside of the immediate aftershock zone. The further one gets from the initiating earthquake in both space and time, the more difficult it is to establish an association.

The physics involved in actually triggering an earthquake is complex. Most earthquake-generating zones are in a state of being close to failure. If such a zone were to be left completely alone, it would generate significant earthquakes spontaneously. Remote earthquakes, however, are in a position to disturb this critical state, either by shifting the stresses statically, or by dynamic change caused by passing seismic waves.

The first type of triggering may be due to static changes in the critical state. For example, after the magnitude 7.3 Landers earthquake struck California in 1992, it is said that "the earthquake map of California lit up like a Christmas tree". This event reinforced the idea of remotely triggered earthquakes, and pushed the hypothesis into the scientific mainstream.

There is scientific evidence for a "long reach", mainly in the form of discrete element modelling used in the mining industry. If rock is modeled as discrete elements in a critical state, a single disturbance can influence a wide area. A smaller-scale example is when a small excavation in a valley triggers a landslide and brings down a whole mountainside.

SLOW EARTHQUAKE

A slow earthquake is a discontinuous, earthquake-like event that releases energy over a period of hours to months, rather than the seconds to minutes characteristic of a typical earthquake. First detected using long term strain measurements, most slow earthquakes now appear to be accompanied by fluid flow and related tremor, which can be detected and approximately located using seismometer data filtered appropriately (typically in the 1–5 Hz band). That is, they are quiet compared to a regular earthquake, but not "silent" as described in the past.

Causes

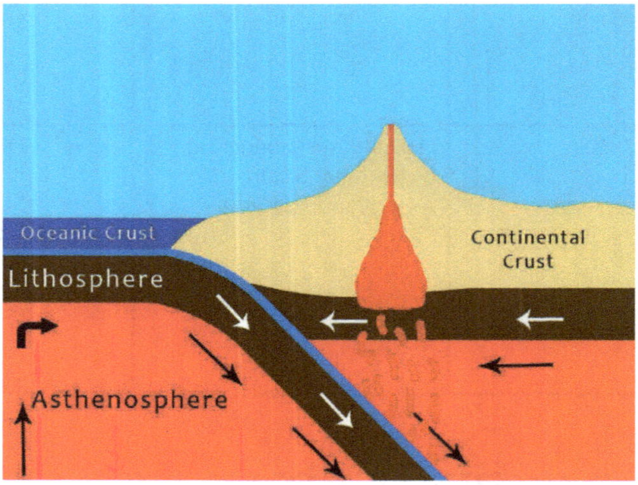

Common Cross Section of a Subduction Zone

Earthquakes occur as a consequence of gradual stress increases in a region, and once it reaches the maximum stress that the rocks can withstand a rupture generates and the resulting earthquake motion is related to a drop in the shear stress of the system. Earthquakes generate seismic waves when the rupture in the system occurs, the seismic waves consist of different types of waves that are capable of moving through the Earth like ripples over water. The causes that lead to slow earthquakes have only been theoretically investigated, by the formation of longitudinal shear cracks that were analyzed using mathematical models. The different distributions of initial stress, sliding frictional stress, and specific fracture energy are all taken into account. If the initial stress minus the sliding frictional stress (with respect to the initial crack) is low, and the specific fracture energy or the strength of the crustal material (relative to the amount of stress) is high then slow earthquakes will occur regularly. In other words, slow earthquakes are caused by a variety of stick-slip and creep processes intermediated between asperity-controlled brittle and ductile fracture. Asperities are tiny bumps and protrusions along the faces of fractures. They are best documented from intermediate crustal levels of certain subduction zones (especially those that dip shallowly — SW Japan, Cascadia, Chile), but appear to occur on other types of faults as well, notably strike-slip plate boundaries such as the San Andreas fault and "mega-landslide" normal faults on the flanks of volcanos.

Locations

Faulting takes place all over Earth; faults can include convergent, divergent, and transform faults, and normally occur on plate margins. As of 2013 some of the locations that have been recently studied for slow earthquakes include: Cascadia,

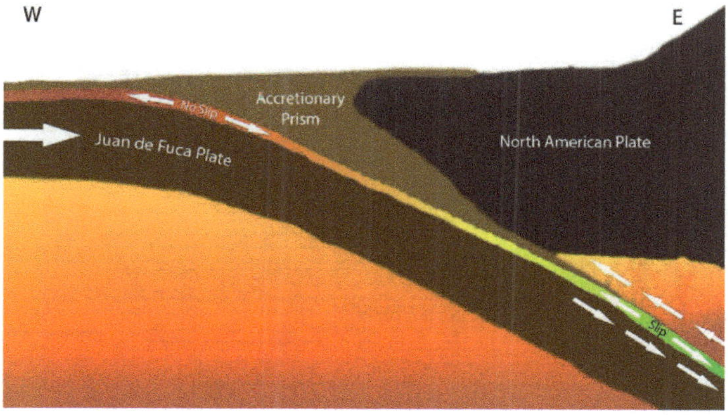

Cascadia Subduction Cross Section

California, Japan, New Zealand, Mexico, and Alaska. The locations of slow earthquakes can provide new insights into the behavior of normal or fast earthquakes. By observing the location of tremors associated with slow-slip and slow earthquakes, seismologists can determine the extension of the system and estimate future earthquakes in the area of study.

Types

Teruyuki Kato identifies various types of slow earthquake:

- low-frequency earthquakes (LFE)
- very low frequency earthquakes (VLF)
- slow-slip events (SSE)
- episodic tremor and slip (ETS)

Episodic Tremor and Slip

Slow earthquakes can be episodic (relative of plate movement), and therefore somewhat predictable, a phenomenon termed "episodic tremor and slip" or "ETS" in the literature. ETS events can last for weeks as opposed to "normal earthquakes" occur in a matter of seconds. Several slow-earthquake events around the world appear to have triggered major, damaging seismic earthquakes in the shallower crust (e.g., 2001 Nisqually, 1995 Antofagasta). Conversely, major earthquakes trigger "post-seismic creep" in the deeper crust and mantle. Just like regular earthquakes, slow earthquakes can cause devastating tsunamis, such as the Mentawai tsunami in 2010. The earthquake that caused this registered a magnitude of 7.8 and struck offshore the Mentawai islands in western Indonesia, causing more than 400 human casualties. Seismologists characterized it as a slow earthquake due to disproportionately large tsunami waves, rupture duration near 125 seconds, shallow near-trench slip, and deficiencies in energy.

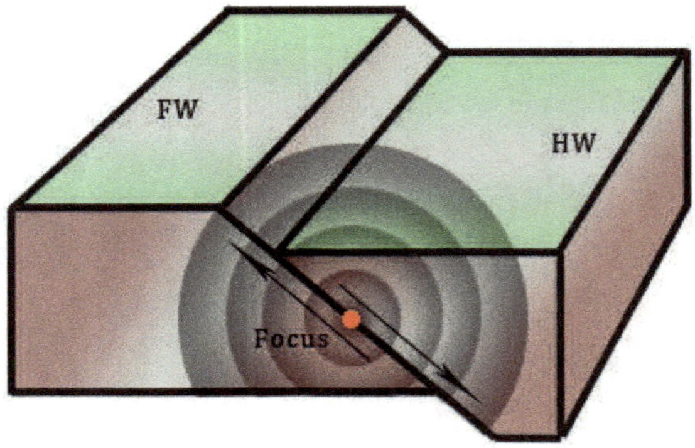

Earthquake FW-HW diagram

Every five years a year-long quake of this type occurs beneath the New Zealand capital, Wellington. It was first measured in 2003, and has reappeared in 2008 and 2013. It lasts for around a year each time, releasing as much energy as a magnitude 7 quake.

SUBMARINE EARTHQUAKE

A submarine, undersea, or underwater earthquake is an earthquake that occurs underwater at the bottom of a body of water, especially an ocean. They are the leading cause of tsunamis. The magnitude can be measured scientifically by the use of the moment magnitude scale and the intensity can be assigned using the Mercalli intensity scale.

Understanding plate tectonics helps to explain the cause of submarine earthquakes. The Earth's surface or lithosphere comprises tectonic plates which average approximately 50 miles in thickness, and are continuously moving very slowly upon a bed of magma in the asthenosphere and inner mantle. The plates converge upon one another, and one subducts below the other, or, where there is only shear stress, move horizontally past each other. Little movements called fault creep are minor and not measurable. The plates meet with each other, and if rough spots cause the movement to stop at the edges, the motion of the plates continue. When the rough spots can no longer hold, the sudden release of the built-up motion releases, and the sudden movement under the sea floor causes a submarine earthquake. This area of slippage both horizontally and vertically is called the epicenter, and has the highest magnitude, and causes the greatest damage.

As with a continental earthquake the severity of the damage is not often caused by the earthquake at the rift zone, but rather by events which are triggered by the earthquake. Where a continental earthquake will cause damage and loss of life on land from fires, damaged structures, and flying objects; a submarine earthquake alters the sea bed floor, resulting in a series of waves, and depending on the length and magnitude of the earthquake, huge tidal waves and tsunamis, which bear down on coastal cities causing property damages and loss of life. Submarine earthquakes can also damage submarine communications cables, leading to widespread disruption of the Internet and international telephone network in those areas. This is particularly common in Asia, where many submarine links cross submarine earthquake zones such as the Pacific Ring of Fire.

Tectonic Plate Boundaries

The different ways in which tectonic plates rub against each other under the ocean or sea floor to create submarine earthquakes. The type of friction created may be due to the characteristic of the geologic fault or the plate boundary as follows. Some of the main areas of large tsunami producing submarine earthquakes are the Pacific Ring of Fire and the Great Sumatran fault.

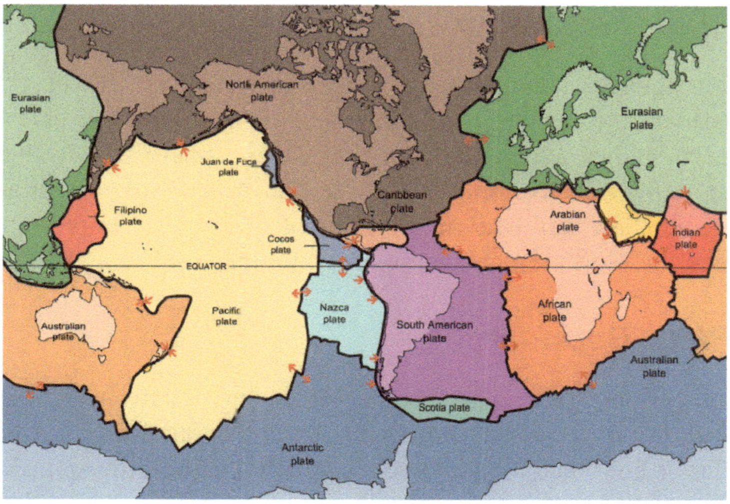

Tectonic plate boundaries, showing the directions of plate movements.

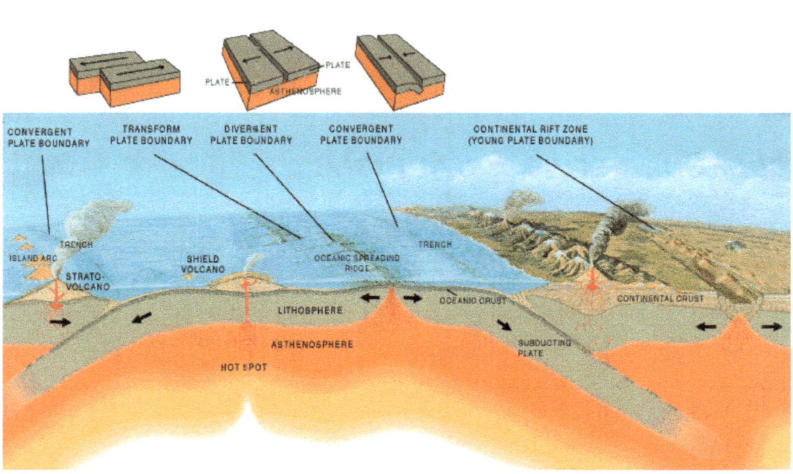

Different kinds of boundaries

Convergent Plate Boundary

The older, and denser plate moves below the lighter plate. The further down it moves, the hotter it becomes, until finally melting altogether at the asthenosphere and inner mantle and the crust is actually destroyed. The location where the two oceanic plates actually meet become deeper and deeper creating trenches with each successive action. There is an interplay of various densities of lithosphere rock, asthenosphere magma, cooling ocean water and plate movement for example the Pacific Ring of Fire. Therefore, the site of the sub oceanic trench will be a site of submarine earthquakes; for example the Mariana Trench, Puerto Rico Trench, and the volcanic arc along the Great Sumatran fault.

Transform Plate Boundary

A transform-fault boundary, or simply a transform boundary is where two plates will slide past each other, and the irregular pattern of their edges may catch on each other. The lithosphere is neither added to from the asthenosphere nor is it destroyed as in convergent plate action. For example, along the San Andreas fault strike-slip fault zone, the Pacific Tectonic Plate has been moving along at about 5 cm/yr in a northwesterly direction, whereas the North American Plate is moving south-easterly.

Divergent Plate Boundary

Rising convection currents occur where two plates are moving away from each other. In the gap, thus produced hot magma rises up, meets the cooler sea water, cools, and solidifies, attaching to either or both tectonic plate edges creating an oceanic spreading ridge. When the fissure again appears, again magma will rise up, and form new lithosphere crust. If the weakness between the two plates allows the heat and pressure of the asthenosphere to build over a large amount of time, a large quantity of magma will be released pusing up on the plate edges and the magma will solidify under the newly raised plate edges, see formation of a submarine volcano. If the fissure is able to come apart because of the two plates moving apart, in a sudden movement, an earthquake tremor may be felt for example at the Mid-Atlantic Ridge between North America and Africa.

List of Major Submarine Earthquakes

The following is a list of major submarine earthquakes.

Date	Event	Location	Estimated moment magnitude	Notes
March 11, 2011	2011 Tōhoku earthquake	The epicenter is 130 kilometers (81 mi) off the east coast of the Oshika Peninsula, Tōhoku, with the hypocenter at a depth of 32 km (20 mi).	9.0	This is strongest known earthquake to hit Japan

December 26, 2006	2006 Hengchun earthquakes	The epicenter is off the southwest coast of Taiwan, in the Luzon Strait, which connects the South China Sea with the Philippine Sea.	7.1	
December 26, 2004	2004 Indian Ocean earthquake	The epicenter is off the northwestern coast of Sumatra, Indonesia.	9.3	This is the third largest earthquake in recorded history and generated massive tsunamis, which caused widespread devastation when they hit land, leaving an estimated 230,000 people dead in countries around the Bay of Bengal and the Indian Ocean.
May 4, 1998		A part of the island of Yonaguni was destroyed by a submarine earthquake.		
May 22, 1960	1960 Valdivia earthquake	The epicenter is off the coast of South Central Chile.	9.5	This is the strongest earthquake ever recorded.
December 20, 1946	1946 Nankaido earthquake	The epicenter is off the southern coast of Kii Peninsula and Shikoku, Japan.	8.1	
December 7, 1944	1944 Tōnankai earthquake	The epicenter is about 20 km off the coast of the Shima Peninsula in Japan.	8.0	
November 18, 1929	1929 Grand Banks earthquake	The epicenter is at Grand Banks, off the south coast of Newfoundland in the Atlantic Ocean.	7.2	
June 15, 1896	1896 Sanriku earthquake	The epicenter is off the Sanriku coast of northeastern Honshū, Japan.	8.5	
April 4, 1771		The epicenter is near Yaeyama Islands in Okinawa, Japan.	7.4	
January 26, 1700	1700 Cascadia earthquake	The epicenter is offshore from Vancouver Island to northern California.	~9.0	This is one of the largest earthquakes on record.

SUPERSHEAR EARTHQUAKE

A supershear earthquake is an earthquake in which the propagation of the rupture along the fault surface occurs at speeds in excess of the seismic shear wave (S-wave) velocity. This causes an effect analogous to a sonic boom.

Rupture Propagation Velocity

During seismic events along a fault surface the displacement initiates at the focus and then propagates outwards. Typically for large earthquakes the focus lies towards one end of the slip surface and much of the propagation is unidirectional (e.g. the 2008 Sichuan and 2004 Indian Ocean earthquakes). Theoretical studies have in the past suggested that the upper bound for propagation velocity is that of Rayleigh waves, approximately 0.92 of the shear wave velocity. However, evidence of propagation at velocities between S-wave and compressional wave (P-wave) values have been reported for several earthquakes in agreement with theoretical and laboratory studies that support the possibility of rupture propagation in this velocity range.

Occurrence

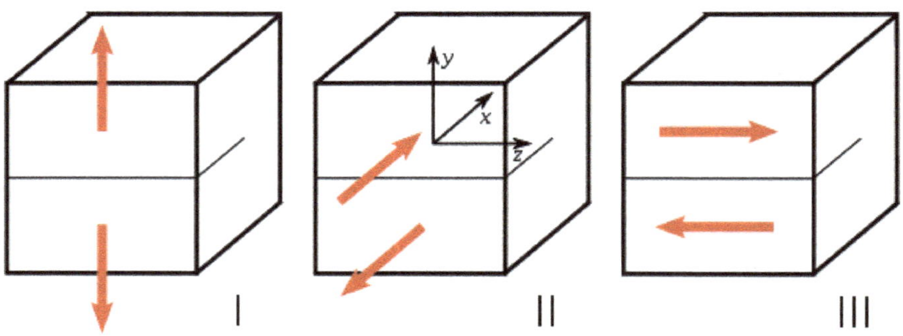

Mode-I, Mode-II, and Mode-III cracks.

Evidence of rupture propagation at velocities greater than S-wave velocities expected for the surrounding crust have been observed for several large earthquakes associated with strike-slip faults. During strike-slip, the main component of rupture propagation will be horizontal, in the direction of displacement, as a Mode II (in-plane) shear crack. This contrasts with a dip-slip rupture where the main direction of rupture propagation will be perpendicular to the displacement, like a Mode III (anti-plane) shear crack. Theoretical studies have shown that Mode III cracks are limited to the shear wave velocity but that Mode II cracks can propagate between the S and P-wave velocities and this may explain why supershear earthquakes have not been observed on dip-slip faults.

Examples

Directly Observed

- 1999 Izmit earthquake, magnitude M_w 7.6 associated with strike-slip movement on the North Anatolian Fault Zone
- 1999 Düzce earthquake, magnitude M_w 7.2 associated with strike-slip movement on the North Anatolian Fault Zone
- 2001 Kunlun earthquake magnitude, M_w 7.8 associated with strike-slip movement on the Kunlun fault
- 2002 Denali earthquake, magnitude M_w 7.9 associated with strike-slip movement on the Denali Fault
- 2010 Yushu earthquake, magnitude M_w 6.9 associated with strike-slip movement on the Yushu Fault
- 2012 Indian Ocean earthquakes, magnitude M_w 8.6 associated with strike-slip on several fault segments - the first supershear event recognised in oceanic lithosphere.
- 2013 Craig earthquakes, magnitude M_w 7.6 associated with strike-slip on the Queen Charlotte Fault - the first supershear event recognised on plate boundaries.

Inferred

- 1906 San Francisco earthquake, magnitude M_w 7.8 associated with strike-slip movement on the San Andreas Fault
- 1979 Imperial Valley earthquake, magnitude M_w 6.4 associated with slip on the Imperial Fault
- 2013 Okhotsk Sea earthquake magnitude M_w 6.7 aftershock was an extremely deep (640 kilometers (400 miles)) supershear as well as unusually fast at "eight kilometers per second (five miles per second), nearly 50 percent faster than the shear wave velocity at that depth."

TSUNAMI EARTHQUAKE

A tsunami earthquake triggers a tsunami of a magnitude that is very much larger than the magnitude of the earthquake as measured by shorter-period seismic waves. The term was introduced by Hiroo Kanamori in 1972. Such events are a result of relatively slow rupture velocities. They are particularly dangerous as a large tsunami may arrive at a neighbouring coast with little or no warning.

Characteristics

The distinguishing feature for a tsunami earthquake is that the release of seismic energy occurs at long periods (low frequencies) relative to typical tsunamigenic earthquakes. Earthquakes of this type do not generally show the peaks of seismic wave activity associated with ordinary events. A tsunami earthquake can be defined as an undersea earthquake for which the surface wave magnitude M_s differs markedly from the moment magnitude M_w, because the former is calculated from surface waves with a period of about 20 seconds, whereas the latter is a measure of the total energy release at all frequencies. The displacements associated with tsunami earthquakes are consistently greater than those associated with ordinary tsunamigenic earthquakes of the same moment magnitude, typically more than double. Rupture velocities for tsunami earthquakes are typically about 1.0 km per second, compared to the more normal 2.5–3.5 km per second for other megathrust earthquakes. These slow rupture speeds lead to greater directivity, with the potential to cause higher run-ups on short coastal sections. Tsunami earthquakes mainly occur at subduction zones where there is a large accretionary wedge or where sediments are being subducted, as this weaker material leads to the slower rupture velocities.

Cause

Analysis of tsunami earthquakes such as the 1946 Aleutian Islands earthquake shows that the release of seismic moment takes place at an unusually long period. Calculations of the effective moment derived from surface waves show a rapid increase with decrease in the frequency of the seismic waves, whereas for ordinary earthquakes it remains almost constant with frequency. The duration over which the seabed is deformed has little effect on the size of the resultant tsunami for times up to several minutes. The observation of long period energy release is consistent with unusually slow rupture propagation velocities. Slow rupture velocities are linked to propagation through relatively weak material, such as poorly consolidated sedimentary rocks. Most tsunami earthquakes have been linked to rupture within the uppermost part of a subduction zone, where an accretionary wedge is developed in the hanging wall of the megathrust. Tsunami earthquakes have also been linked to the presence of a thin layer of subducted sedimentary rock along the uppermost part of the plate interface, as is thought to be present in areas of significant topography at the top of the oceanic crust, and where propagation was in an up-dip direction, possibly reaching the seafloor.

Identifying Tsunami Earthquakes

Standard methods of giving early warnings for tsunamis rely on data that will not typically identify a tsunami earthquake as tsunamigenic and therefore fail to predict possibly damaging tsunamis.

Examples

1896 Sanriku

On 15 June 1896 the Sanriku coast was struck by a devastating tsunami with a maximum wave height of 38.2 m, which caused more than 22,000 deaths. The residents of the coastal towns and villages were taken completely by surprise because the tsunami had only been preceded by a relatively weak shock. The magnitude of the tsunami has been estimated as M_t=8.2 while the earthquake shaking only indicated a magnitude of M_s=7.2.

This discrepancy in magnitude requires more than just a slow rupture velocity. Modelling of tsunami generation that takes into account additional uplift associated with deformation of the softer sediments of the accretionary wedge caused by horizontal movement of the 'backstop' in the overriding plate has successfully explained the discrepancy, estimating a magnitude of M_w=8.0–8.1.

1992 Nicaragua

Main article: 1992 Nicaragua earthquake

The 1992 Nicaragua earthquake was the first tsunami earthquake to be recorded with a broad-band seismic network.

Other Tsunami Earthquakes

- 1605 Keichō Nankaidō earthquake
- 1946 Aleutian Islands earthquake
- November 1960 Peru earthquake
- 1963 Kuril Islands earthquake
- 1975 Kuril Islands earthquake
- 1994 Java earthquake
- 1996 Chimbote earthquake
- 2006 Pangandaran earthquake and tsunami

REFERENCES

1. Gates, A.; Ritchie, D. (2006). Encyclopedia of Earthquakes and Volcanoes. Infobase Publishing. p. 89. ISBN 978-0-8160-6302-4. Retrieved 29 November 2010.
2. National Research Council (U.S.). Committee on the Science of Earthquakes (2003). "5. Earthquake Physics and Fault-System Science". Living on an Active Earth: Perspectives on Earthquake Science. Washington D.C.: National Academies Press. p. 418. ISBN 978-0-309-06562-7. Retrieved 29 November 2010.
3. Kayal, J.R. (2008). Microearthquake seismology and seismotectonics of South Asia. Springer. p. 15. ISBN 978-1-4020-8179-8. Retrieved 29 November 2010.
4. Vallée, M. (2007). "Rupture Properties of the Giant Sumatra Earthquake Imaged by Empirical Green's Function Analysis" (PDF). Bulletin of the Seismological Society of America. 97 (1A): S103–S114. Bibcode:2007BuSSA..97S.103V. doi:10.1785/0120050616. Retrieved 29 November 2010.
5. Maeda, K. (1999). "Time distribution of immediate foreshocks obtained by a stacking method". In Wyss M., Shimazaki K. & Ito A. Seismicity patterns, their statistical significance and physical meaning. Reprint from Pageoph Topical Volumes. Birkhäuser. pp. 381–394. ISBN 978-3-7643-6209-6. Retrieved 29 November 2010.

6. Felzer, K.R.; Abercrombie R.E.; Ekström G. (2004). "A Common Origin for Aftershocks, Foreshocks, and Multiplets" (PDF). Bulletin of the Seismological Society of America. 94 (1). Bibcode:2004BuSSA..94...88F. doi:10.1785/0120030069. Retrieved 29 November 2010.
7. Ludwin, R. (16 September 2004). "Earthquake Prediction". The Pacific Northwest Seismic Network. Retrieved 29 November 2010.
8. McGuire, J.J.; Boettcher M.S.; Jordan T.H. (2005). "Foreshock sequences and short-term earthquake predictability on East Pacific Rise transform faults". Nature. 434 (7032): 457–461. Bibcode:2005Natur.434..457M. doi:10.1038/nature03377. PMID 15791246. Retrieved 29 November 2010.
9. "El Sismo del 20 de Octubre de 2006" (PDF) (in Spanish). IGP.
10. "Informe de sismo sensible" (in Spanish). GUC.
11. "Magnitude 7.3 - NEAR THE EAST COAST OF HONSHU, JAPAN". USGS.
12. "Informe de sismo sensible" (in Spanish). GUC.
13. Omori, F. (1894). "On the aftershocks of earthquakes" (PDF). Journal of the College of Science, Imperial University of Tokyo. 7: 111–200.
14. Utsu, T. (1961). "A statistical study of the occurrence of aftershocks". Geophysical Magazine. 30: 521–605.
15. Utsu, T.; Ogata, Y.; Matsu'ura, R.S. (1995). "The centenary of the Omori formula for a decay law of aftershock activity" (PDF). Journal of Physics of the Earth. 43: 1–33. Archived from the original (PDF) on 2015-07-16.
16. Quigley, M. "New Science update on 2011 Christchirch Earthquake for press and public: Seismic fearmongering or time to jump ship". Christchurch Earthquake Journal. Archived from the original on 29 January 2012. Retrieved 25 January 2012.
17. Richter, Charles F., Elementary seismology (San Francisco, California, USA: W. H. Freeman & Co., 1958), page 69.
18. Båth, Markus (1965). "Lateral inhomogeneities in the upper mantle". Tectonophysics. 2: 483–514. Bibcode:1965Tectp...2..483B. doi:10.1016/0040-1951(65)90003-X.
19. Elizabeth K. Gardner (2009-03-13). "New Madrid fault system may be shutting down". physorg.com. Retrieved 2011-03-25.
20. Wallace, Robert E. "Present-Day Crustal Movements and the Mechanics of Cyclic Deformation". The San Andreas Fault System, California. Archived from the original on 2006-12-16. Retrieved 2007-10-26.
21. "Earthquakes Actually Aftershocks Of 19th Century Quakes; Repercussions Of 1811 And 1812 New Madrid Quakes Continue To Be Felt". Science Daily. Archived from the original on 8 November 2009. Retrieved 2009-11-04.
22. McGuire JJ, Boettcher MS, Jordan TH (2005). "Foreshock sequences and short-term earthquake predictability on East Pacific Rise transform faults". Nature. 434 (7032): 445–7. Bibcode:2005Natur.434..457M. doi:10.1038/nature03377. PMID 15791246.

23. For example: Helmstetter, Agnès; Sornette, Didier (October 2003). "Predictability in the Epidemic-Type Aftershock Sequence model of interacting triggered seismicity". Journal of Geophysical Research: Solid Earth. 108 (B10): 2482ff. arXiv:cond-mat/0208597Freely accessible. Bibcode:2003JGRB..108.2482H. doi:10.1029/2003JB002485. Retrieved 2013-04-10. As part of an effort to develop a systematic methodology for earthquake forecasting, we use a simple model of seismicity based on interacting events which may trigger a cascade of earthquakes, known as the Epidemic-Type Aftershock Sequence model (ETAS).
24. Japanese researchers diagnose hundreds of cases of 'earthquake sickness', Daily Telegraph, 20 June 2016
25. After the earthquake: why the brain gives phantom quakes, The Guardian, 6 November 2016
26. "Magnitude 7.8 - VANUATU #Summary". USGS. 2009-10-07. Retrieved 2009-10-09.
27. "How One Big Earthquake Triggers Another". Live Science. Retrieved 2016-10-26.
28. "Doublet Earthquakes And Earthquake Dynamics". scientificblogging.com. January 30, 2008. Retrieved 2010-02-03.
29. "Doublet Earthquakes And Earthquake Dynamics". Science 2.0. 2014-08-27. Retrieved 2016-11-01.
30. "Central Kuril Island Tsunami in Crescent City and California.html". www.usc.edu. Retrieved 2016-11-01.

CHAPTER 3

Causes of Earthquake

Chapter Contents

Most earthquakes are causally related to compressional or tensional stresses built up at the margins of the huge moving lithospheric plates that make up the earth's surface. The immediate cause of most shallow earthquakes is the sudden release of stress along a fault, or fracture in the earth's crust, resulting in movement of the opposing blocks of rock past one another. These movements cause vibrations to pass through and around the earth in wave form, just as ripples are generated when a pebble is dropped into water. Volcanic eruptions, rockfalls, landslides, and explosions can also cause a quake, but most of these are of only local extent. Shock waves from a powerful earthquake can trigger smaller earthquakes in a distant location hundreds of miles away if the geologic conditions are favorable.

INTRODUCTION

Earthquakes are one of the most dangerous natural disasters on Earth. This is because they strike with little or no warning, and can cause catastrophic damage. All that shaking comes from deep underground, but as you know, the surface shakes a lot too, which is where all the damage occurs. Buildings fall down, roads and bridges collapse, and land and mud come sliding down from hillsides.

But what causes all that shaking in the first place? Earthquakes happen deep underground along tectonic plate boundaries. Tectonic plates are what make up the earth's crust, its outermost layer. These plates fit together like puzzle pieces but they don't stay in one place. They're always moving because the part of the earth underneath them is like a fluid. And because the plates are sitting on top of this fluid like ice on top of a pond, they are not locked in place and are sort of floating about.

However, each plate is lined up pretty well with the other plates around it. So as they move, they create tension and pressure as they slide past and bump into each other, sometimes even sticking together. And though the plate boundary is stuck, the plate itself keeps moving and pulling the rest of the plate with it. Eventually, the pulling becomes too much and the plates suddenly break free from each other, causing an earthquake.

FAULT MOVEMENT

In geology, a fault is a planar fracture or discontinuity in a volume of rock, across which there has been significant displacement as a result of rock mass movement. Large faults within the Earth's crust result from the action of plate tectonic forces, with the largest forming the boundaries between the plates, such as subduction zones or transform faults. Energy release associated with rapid movement on active faults is the cause of most earthquakes.

A *fault plane* is the plane that represents the fracture surface of a fault. A *fault trace* or *fault line* is the intersection of a fault plane with the ground surface. A fault trace is also the line commonly plotted on geologic maps to represent a fault.

Since faults do not usually consist of a single, clean fracture, geologists use the term *fault zone* when referring to the zone of complex deformation associated with the fault plane.

The two sides of a non-vertical fault are known as the *hanging wall* and *footwall*. By definition, the hanging wall occurs above the fault plane and the footwall occurs below the fault. This terminology comes from mining: when working a tabular ore body, the miner stood with the footwall under his feet and with the hanging wall above him.

Mechanisms of Faulting

Because of friction and the rigidity of rocks, they cannot glide or flow past each other easily, and occasionally all movement stops. When this happens, stress builds up in rocks and when it reaches a level that exceeds the strain threshold, the accumulated potential energy is dissipated by the release of strain, which is focused into a plane along which relative motion is accommodated — the fault.

Normal fault in La Herradura Formation, Morro Solar, Peru. Note the relative displacement of the bright horizon. At closer inspection a second normal fault can be seen to the right.

Strain occurs accumulatively or instantaneously, depending on the rheology of the rock; the ductile lower crust and mantle accumulates deformation gradually via shearing, whereas the brittle upper crust reacts by fracture - instantaneous stress release - to cause motion along the fault. A fault in ductile rocks can also release instantaneously when the strain rate is too great. The energy released by instantaneous strain-release causes earthquakes, a common phenomenon along transform boundaries.

Slip, Heave, Throw

Slip is defined as the relative movement of geological features present on either side of a fault plane, and is a displacement vector.

A fault in Morocco.The fault plane is the steeply leftward-dipping line in the centre of the photo, which is the plane along which the rock layers to the left have slipped downwards, relative to the layers to the right of the fault.

A fault's *sense of slip* is defined as the relative motion of the rock on each side of the fault with respect to the other side. In measuring the horizontal or vertical separation, the *throw* of the fault is the vertical component of the dip separation and the *heave* of the fault is the horizontal component, as in "throw up and heave out".

Microfault showing a piercing point (the coin's diameter is 18 mm)

The vector of slip can be qualitatively assessed by studying any drag folding of strata, which may be visible on either side of the fault; the direction and magnitude of heave and throw can be measured only by finding common intersection points on either side of the fault (called a piercing point). In practice, it is usually only possible to find the slip direction of faults, and an approximation of the heave and throw vector.

Fault Types

Based on direction of slip, faults can be generally categorized as:

- *strike-slip*, where the offset is predominantly horizontal, parallel to the fault trace.
- *dip-slip*, offset is predominantly vertical and/or perpendicular to the fault trace.
- *oblique-slip*, combining significant strike and dip slip.

Strike-Slip Faults

The fault surface is usually near vertical and the footwall moves either left or right or laterally with very little vertical motion. Strike-slip faults with left-lateral motion are also known as *sinistral* faults. Those with right-lateral motion are also known as *dextral* faults. Each is defined by the direction of movement of the ground on the opposite side of the fault from an observer.

The Piqiang Fault, a northwest trending left-lateral strike-slip fault in the Taklamakan Desert south of the Tien Shan Mountains, China (40.3°N, 77.7°E)

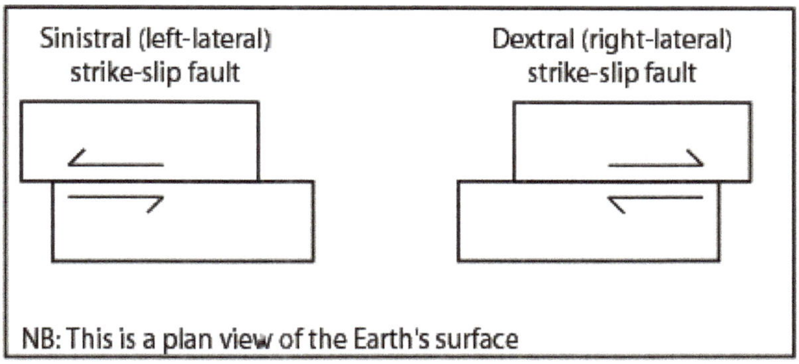

Schematic illustration of the two strike-slip fault types.

A special class of strike-slip faults is the transform fault, where such faults form a plate boundary. These are found related to offsets in spreading centers, such as mid-ocean ridges, and less commonly within continental lithosphere, such as the Dead Sea Transform in the Middle East, or the Alpine Fault, New Zealand. Transform faults are also referred to as conservative plate boundaries, as lithosphere is neither created nor destroyed.

Dip-Slip Faults

Dip-slip faults can occur either as "reverse" or as "normal" faults. A normal fault occurs when the crust is extended. Alternatively such a fault can be called an extensional fault.

Normal faults in Spain, between which rock layers have slipped downwards (at photo's Centre)

The hanging wall moves downward, relative to the footwall. A downthrown block between two normal faults dipping towards each other is called a graben. An upthrown block between two normal faults dipping away from each other is called a horst. Low-angle normal faults with regional tectonic significance may be designated detachment faults.

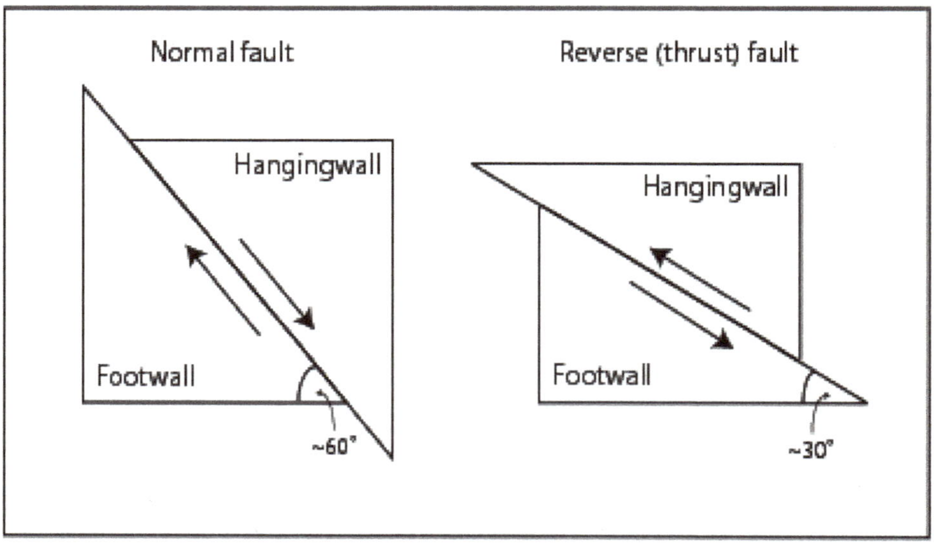

Cross-sectional illustration of normal and reverse dip-slip faults

A reverse fault is the opposite of a normal fault—the hanging wall moves up relative to the footwall. Reverse faults indicate compressive shortening of the crust. The dip of a reverse fault is relatively steep, greater than 45°.

A thrust fault has the same sense of motion as a reverse fault, but with the dip of the fault plane at less than 45°. Thrust faults typically form ramps, flats and fault-bend (hanging wall and foot wall) folds.

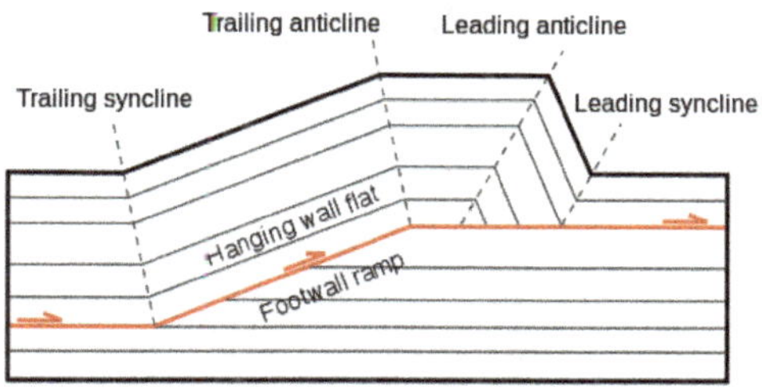

Fault-bend fold

Flat segments of thrust fault planes are known as *flats*, and inclined sections of the thrust are known as *ramps*. Typically, thrust faults move *within* formations by forming flats, and climb up section with ramps.

Fault-bend folds are formed by movement of the hanging wall over a non-planar fault surface and are found associated with both extensional and thrust faults.

Faults may be reactivated at a later time with the movement in the opposite direction to the original movement (fault inversion). A normal fault may therefore become a reverse fault and vice versa.

Thrust faults form nappes and klippen in the large thrust belts. Subduction zones are a special class of thrusts that form the largest faults on Earth and give rise to the largest earthquakes.

Oblique-Slip Faults

A fault which has a component of dip-slip and a component of strike-slip is termed an *oblique-slip fault*. Nearly all faults will have some component of both dip-slip and strike-slip, so defining a fault as oblique requires both dip and strike components to be measurable and significant. Some oblique faults occur within transtensional and transpressional regimes, others occur where the direction of extension or shortening changes during the deformation but the earlier formed faults remain active.

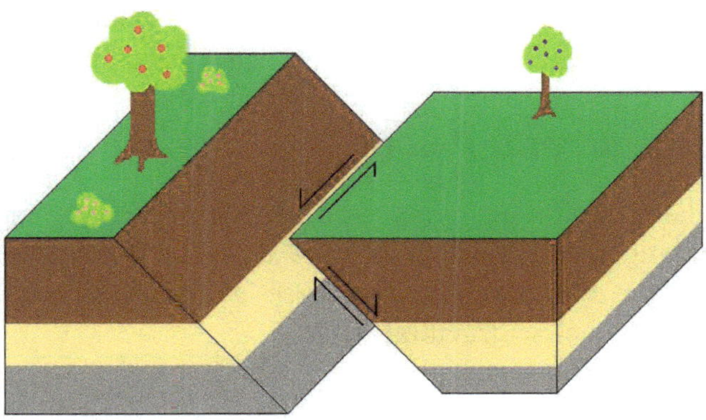

Oblique-slip fault: Arrows represent relative movement.

Oblique-slip fault

The *hade* angle is defined as the complement of the dip angle; it is the angle between the fault plane and a vertical plane that strikes parallel to the fault.

Listric Fault

Listric faults are similar to normal faults but the fault plane curves, the dip being steeper near the surface, then shallower with increased depth.

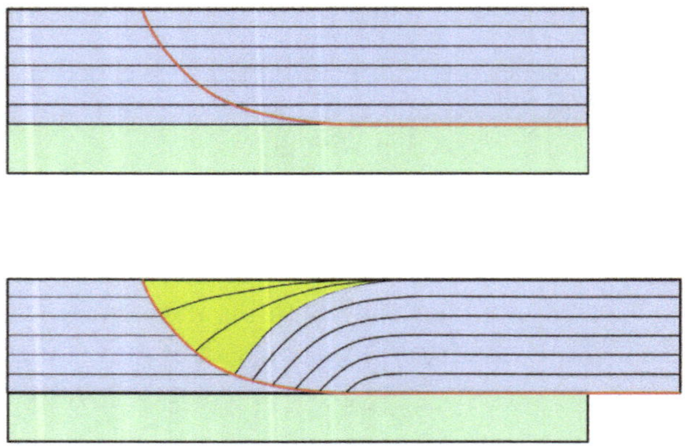

Listric fault (red line)

The dip may flatten into a sub-horizontal décollement, resulting in horizontal slip on a horizontal plane. The illustration shows slumping of the hanging wall along a listric fault. Where the hanging wall is absent (such as on a cliff) the footwall may slump in a manner that creates multiple listric faults.

Ring Fault

Ring faults are faults that occur within collapsed volcanic calderas and the sites of bolide strikes, such as the Chesapeake Bay impact crater. Ring faults may be filled by ring dikes.

Synthetic and Antithetic Faults

Synthetic and antithetic faults are terms used to describe minor faults associated with a major fault. Synthetic faults dip in the same direction as the major fault while the antithetic faults dip in the opposite direction. These faults may be accompanied by rollover anticlines (e.g. the Niger Delta Structural Style).

Fault Rock

All faults have a measurable thickness, made up of deformed rock characteristic of the level in the crust where the faulting happened, of the rock types affected by the fault and of the presence and nature of any mineralizing fluids. Fault rocks are classified by their textures and the implied mechanism of deformation.

Salmon-colored fault gouge and associated fault separates two different rock types on the left (dark grey) and right (light grey). From the Gobi of Mongolia.

Inactive fault from Sudbury to Sault Ste. Marie, Northern Ontario, Canada

A fault that passes through different levels of the lithosphere will have many different types of fault rock developed along its surface. Continued dip-slip displacement tends to juxtapose fault rocks characteristic of different crustal levels, with varying degrees of overprinting. This effect is particularly clear in the case of detachment faults and major thrust faults.

The main types of fault rock include:

- Cataclasite - a fault rock which is cohesive with a poorly developed or absent planar fabric, or which is incohesive, characterised by generally angular clasts and rock fragments in a finer-grained matrix of similar composition.
    - Tectonic or Fault breccia - a medium- to coarse-grained cataclasite containing >30% visible fragments.
    - Fault gouge - an incohesive, clay-rich fine- to ultrafine-grained cataclasite, which may possess a planar fabric and containing <30% visible fragments. Rock clasts may be present
        - Clay smear - clay-rich fault gouge formed in sedimentary sequences containing clay-rich layers which are strongly deformed and sheared into the fault gouge.
- Mylonite - a fault rock which is cohesive and characterized by a well-developed planar fabric resulting from tectonic reduction of grain size, and commonly containing rounded porphyroclasts and rock fragments of similar composition to minerals in the matrix
- Pseudotachylite - ultrafine-grained glassy-looking material, usually black and flinty in appearance, occurring as thin planar veins, injection veins or as a matrix to pseudoconglomerates or breccias, which infills dilation fractures in the host rock.

Impacts on Structures and People

In geotechnical engineering a fault often forms a discontinuity that may have a large influence on the mechanical behavior (strength, deformation, etc.) of soil and rock masses in, for example, tunnel, foundation, or slope construction.

The level of a fault's activity can be critical for (1) locating buildings, tanks, and pipelines and (2) assessing the seismic shaking and tsunami hazard to infrastructure and people in the vicinity. In California, for example, new building construction has been prohibited directly on or near faults that have moved within the Holocene Epoch (the last 11,700 years) of the Earth's geological history. Also, faults that have shown movement during the Holocene plus Pleistocene Epochs (the last 2.6 million years) may receive consideration, especially for critical structures such as power plants, dams, hospitals, and schools. Geologists assess a fault's age by studying soil features seen in shallow excavations and geomorphology seen in aerial photographs. Subsurface clues include shears and their relationships to carbonate nodules, eroded clay, and iron oxide mineralization, in the case of older soil, and lack of such signs in the case of younger soil. Radiocarbon dating of organic material buried next to or over a fault shear is often critical in distinguishing active from inactive faults. From such relationships, paleoseismologists can estimate the sizes of past earthquakes over the past several hundred years, and develop rough projections of future fault activity.

INDUCED SEISMICITY

Induced seismicity refers to typically minor earthquakes and tremors that are caused by human activity that alters the stresses and strains on the Earth's crust. Most induced seismicity is of a low magnitude. A few sites regularly have larger quakes, such as The Geysers geothermal plant in California which averaged two M4 events and 15 M3 events every year from 2004 to 2009. Results of ongoing multi-year research on induced earthquakes by the United States Geological Survey (USGS) published in 2015 suggested that most of the significant earthquakes in Oklahoma, such as the 1952 magnitude 5.7 El Reno earthquake may have been induced by deep injection of waste water by the oil industry. "Earthquake rates have recently increased markedly in multiple areas of the Central and Eastern United States (CEUS), especially since 2010, and scientific studies have linked the majority of this increased activity to wastewater injection in deep disposal wells.

Causes

There are many ways in which induced seismicity has been seen to occur. In the past several years, some energy technologies that inject or extract fluid from the Earth, such as oil and gas extraction and geothermal energy development, have been found or suspected to cause seismic events.

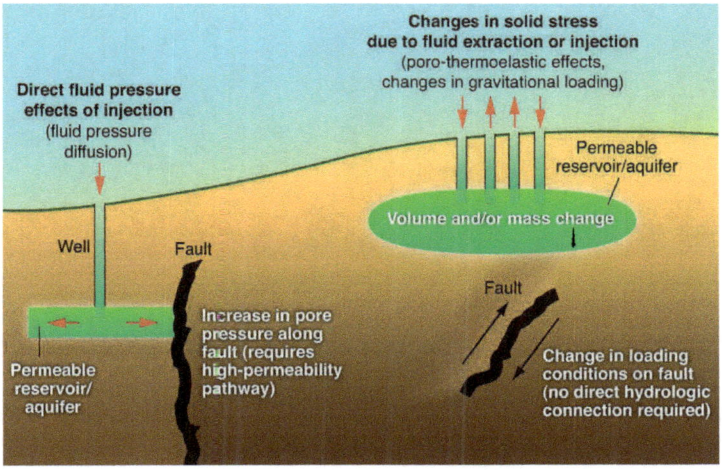

Diagram showing the effects that fluid injection and withdrawal can have on nearby faults can cause induced seismicity.

Some energy technologies also produce wastes that may be managed through disposal or storage by injection deep into the ground. For example, waste water from oil and gas production and carbon dioxide from a variety of industrial processes may be managed through underground injection.

Artificial Lakes

The column of water in a large and deep artificial lake alters in-situ stress along an existing fault or fracture. In these reservoirs, the weight of the water column can significantly change the stress on an underlying fault or fracture by increasing the total stress through direct loading, or decreasing the effective stress through the increased pore water pressure. This significant change in stress can lead to sudden movement along the fault or fracture, resulting in an earthquake. Reservoir-induced seismic events can be relatively large compared to other forms of induced seismicity. Though understanding of reservoir-induced seismic activity is very limited, it has been noted that seismicity appears to occur on dams with heights larger than 330 feet (100 m). The extra water pressure created by large reservoirs is the most accepted explanation for the seismic activity. When the reservoirs are filled or drained, induced seismicity can occur immediately or with a small time lag.

The first case of reservoir-induced seismicity occurred in 1932 in Algeria's Oued Fodda Dam.

The largest earthquake attributed to reservoir-induced seismicity occurred at Koyna Dam

The 6.3 magnitude 1967 Koynanagar earthquake occurred in Maharashtra, India with its epicenter, fore- and aftershocks all located near or under the Koyna Dam reservoir. 180 people died and 1,500 were left injured. The effects of the earthquake were felt 140 mi (230 km) away in Bombay with tremors and power outages.

During the beginnings of the Vajont Dam in Italy, there were seismic shocks recorded during its initial fill. After a landslide almost filled the reservoir in 1963, causing a massive flooding and around 2,000 deaths, it was drained and consequently seismic activity was almost non-existent.

On August 1, 1975, a magnitude 6.1 earthquake at Oroville, California, was attributed to seismicity from a large earth-fill dam and reservoir recently constructed and filled.

The filling of the Katse Dam in Lesotho, and the Nurek Dam in Tajikistan is an example. In Zambia, Kariba Lake may have provoked similar effects.

The 2008 Sichuan earthquake, which caused approximately 68,000 deaths, is another possible example. An article in *Science* suggested that the construction and filling of the Zipingpu Dam may have triggered the earthquake.

Some experts worry that the Three Gorges Dam in China may cause an increase in the frequency and intensity of earthquakes.

Mining

Mining leaves voids that generally alter the balance of forces in the rock, many times causing rock bursts. These voids may collapse producing seismic waves and in some cases reactivate existing faults causing minor earthquakes. Natural cavern collapse forming sinkholes would produce an essentially identical local *seismic event.*

Waste Disposal Wells

Injecting liquids into waste disposal wells, most commonly in disposing of produced water from oil and natural gas wells, has been known to cause earthquakes. This high-saline water is usually pumped into salt water disposal (SWD) wells. The resulting increase in subsurface pore pressure can trigger movement along faults, resulting in earthquakes.

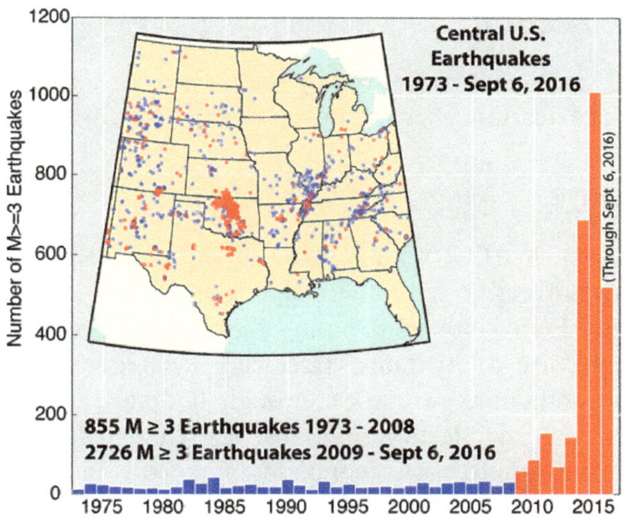

Cumulative number of earthquakes in the central U.S. The red cluster at the center of the map shows an area in and around Oklahoma which experienced the largest increase in activity since 2009.

The 2011 Oklahoma earthquake near Prague, of magnitude 5.8, occurred after 20 years of injecting waste water into porous deep formations at increasing pressures and saturation. On September 3, 2016, an even stronger earthquake with a magnitude of 5.8 occurred near Pawnee, Oklahoma, followed by nine aftershocks between magnitudes 2.6 and 3.6 within 3 1/2 hours. Tremors were felt as far away as Memphis, Tennessee, and

Gilbert, Arizona. Mary Fallin, the Oklahoma governor, declared a local emergency and shutdown orders for local disposal wells were ordered by the Oklahoma Corporation Commission. Results of ongoing multi-year research on induced earthquakes by the United States Geological Survey (USGS) published in 2015 suggested that most of the significant earthquakes in Oklahoma, such as the 1952 magnitude 5.5 El Reno earthquake may have been induced by deep injection of waste water by the oil industry. Prior to April 2015 however, the Oklahoma Geological Survey's position was that the quake was most likely due to natural causes and was not triggered by waste injection. This was one of many earthquakes which have affected the Oklahoma region.

Since 2009 earthquakes have become hundreds of time more common in Oklahoma with magnitude 3 events increasing from 1 or 2 per year to 1 or 2 per day. On April 21, 2015, the Oklahoma Geological Survey released a statement reversing its stance on induced earthquakes in Oklahoma: "The OGS considers it very likely that the majority of recent earthquakes, particularly those in central and north-central Oklahoma, are triggered by the injection of produced water in disposal wells."

Extraction of Fossil Fuels

Large-scale fossil fuel extraction can generate earthquakes.

Groundwater Extraction

The changes in crustal stress patterns caused by the large scale extraction of groundwater has been shown to trigger earthquakes, as in the case of the 2011 Lorca earthquake.

Geothermal Energy

Enhanced geothermal systems (EGS), a new type of geothermal power technologies that do not require natural convective hydrothermal resources, are known to be associated with induced seismicity. EGS involves pumping fluids at pressure to enhance or create permeability through the use of hydraulic fracturing techniques. Hot dry rock (HDR) EGS actively creates geothermal resources through hydraulic stimulation. Depending on the rock properties, and on injection pressures and fluid volume, the reservoir rock may respond with tensile failure, as is common in the oil and gas industry, or with shear failure of the rock's existing joint set, as is thought to be the main mechanism of reservoir growth in EGS efforts.

HDR and EGS systems are currently being developed and tested in Soultz-sous-Forêts (France), Desert Peak and the Geysers (U.S.), Landau (Germany), and Paralana and Cooper Basin (Australia). Induced seismicity events at the Geysers geothermal field in California has been strongly correlated with injection data. The test site at Basel, Switzerland, has been shut down due to induced seismic events.

Largest Events at EGS Sites Worldwide	
Site	**Maximum Magnitude**
Cerro Prieto, Baja California, Mexico	6.6

The Geysers, United States	4.6
Cooper Basin, Australia	3.7
Basel, Switzerland	3.4
Rosemanowes Quarry, United Kingdom	3.1
Soultz-sous-Forêts, France	2.9

Researchers at MIT believe that seismicity associated with hydraulic stimulation can be mitigated and controlled through predictive siting and other techniques. With appropriate management, the number and magnitude of induced seismic events can be decreased, significantly reducing the probability of a damaging seismic event.

Induced seismicity in Basel led to suspension of its HDR project. A seismic hazard evaluation was then conducted, which resulted in the cancellation of the project in December 2009.

Hydraulic Fracturing

Hydraulic fracturing is a technique in which high-pressure fluid is injected into the low-permeable reservoir rocks in order to induce fractures to increase hydrocarbon production. This process usually generates seismic events that are too small to be felt at the surface (with magnitudes ranging from -3 to 0), although several cases of larger magnitude events (M > 4) have been recorded in Canada in the unconventional resources of Alberta and British Columbia.

U.S. National Research Council Report

A 2012 report from the U.S. National Research Council examined the potential for energy technologies—including shale gas recovery, carbon capture and storage, geothermal energy production, and conventional oil and gas development—to cause earthquakes. The report found that only a very small fraction of injection and extraction activities among the hundreds of thousands of energy development sites in the United States have induced seismicity at levels noticeable to the public. However, although scientists understand the general mechanisms that induce seismic events, they are unable to accurately predict the magnitude or occurrence of these earthquakes due to insufficient information about the natural rock systems and a lack of validated predictive models at specific energy development sites.

The report noted that hydraulic fracturing has a low risk for inducing earthquakes that can be felt by people, but underground injection of wastewater produced by hydraulic fracturing and other energy technologies has a higher risk of causing such earthquakes. In addition, carbon capture and storage—a technology for storing excess carbon dioxide underground—may have the potential for inducing seismic events, because significant volumes of fluids are injected underground over long periods of time.

REFERENCES

1. Brodie, Kate; Fettes, Douglas; Harte, Ben; Schmid, Rolf (29 January 2007), Structural terms including fault rock terms, International Union of Geological Sciences
2. Davis, George H.; Reynolds, Stephen J. (1996), "Folds", Structural Geology of Rocks and Regions (2nd ed.), John Wiley & Sons, pp. 372–424, ISBN 0-471-52621-5
3. Fichter, Lynn S.; Baedke, Steve J. (13 September 2000), A Primer on Appalachian Structural Geology, James Madison University, retrieved 19 March 2010
4. Hart, E.W.; Bryant, W.A., (1997), Fault rupture hazard in California: Alquist-Priolo earthquake fault zoning act with index to earthquake fault zone maps, Special Publication 42., California Division of Mines and Geology
5. Marquis, John; Hafner, Katrin; Hauksson, Egill, "The Properties of Fault Slip", Investigating Earthquakes through Regional Seismicity, Southern California Earthquake Center, retrieved 19 March 2010
6. McKnight, Tom L.; Hess, Darrel (2000), "The Internal Processes: Types of Faults", Physical Geography: A Landscape Appreciation, Prentice Hall, pp. 416–7, ISBN 0-13-020263-0
7. Park, R.G. (2004), Foundation of Structural Geology (3 ed.), Routledge, p. 11, ISBN 978-0-7487-5802-9
8. Tingley, J.V.; Pizarro, K.A. (2000), Traveling America's loneliest road: a geologic and natural history tour, Nevada Bureau of Mines and Geology Special Publication, 26, Nevada Bureau of Mines and Geology, p. 132, ISBN 978-1-888035-05-6, retrieved 2010-04-02
9. USGS, Hanging wall Foot wall, retrieved 2 April 2010
10. USGS, Earthquake Glossary - fault trace, retrieved 10 April 2015
11. USGS (30 April 2003), Where are the Fault Lines in the United States East of the Rocky Mountains?, retrieved 6 March 2010
12. "Man-made geothermal earthquakes". Anderson Springs Community Alliance. 2009. Retrieved April 28, 2016.
13. Hough, Susan E.; Page, Morgan (October 20, 2015). "A Century of Induced Earthquakes in Oklahoma?". U.S. Geological Survey. Retrieved November 8, 2015. Several lines of evidence further suggest that most of the significant earthquakes in Oklahoma during the 20th century may also have been induced by oil production activities. Deep injection of waste water, now recognized to potentially induce earthquakes, in fact began in the state in the 1930s.
14. Ellsworth, W.L. (2013). "Injection-induced earthquakes". Science. 341 (6142): 7. doi:10.1126/science.1225942.
15. Keranen, K.M.; Weingarten, Matthew; Abers, G.A.; Bekins, B.A.; Ge, Shemin (2014). "Sharp increase in central Oklahoma seismicity since 2008 induced by massive wastewater injection". Science. 345 (6195): 448–451. doi:10.1126/

science.1255802.

16. Walsh, F.R.; Zoback, M.D. (2015). "Oklahoma's recent earthquakes and saltwater disposal". Science Advances. 1 (5): e1500195. doi:10.1126/sciadv.1500195.
17. Weingarten, Matthew; Ge, Shemin; Godt, J.W.; Bekins, B.A.; Rubinstein, J.L. (2015). "High-rate injection is associated with the increase in U.S. mid-continent seismicity". Science. 348 (6241): 1336–1340. doi:10.1126/science.aab1345.
18. Petersen, Mark D.; Mueller, Charles S.; Moschetti, Morgan P.; Hoover, Susan M.; Llenos, Andrea L.; Ellsworth, William L.; Michael, Andrew J.; Rubinstein, Justin L.; McGarr, Arthur F.; Rukstales, Kenneth S. (April 1, 2016). 2016 One-Year Seismic Hazard Forecast for the Central and Eastern United States from Induced and Natural Earthquakes (PDF) (Report). Reston, Virginia. p. 58. doi:10.3133/ofr20161035. ISSN 2331-1258.
19. Keranen, Katie M.; Savage, Heather M.; Abers, Geoffrey A.; Cochran, Elizabeth S. (2013). "Potentially induced earthquakes in Oklahoma, USA: Links between wastewater injection and the 2011 Mw 5.7 earthquake sequence". Geology. 41: 699–702. doi:10.1130/G34045.1. Retrieved April 28, 2016.via EBSCO
20. Simpson, D. W.; Leith, W. S.; Schcolz, C.H. (1988). "Two Types of Reservoir-Induced Seismicity". Bulletin of the Seismological Society of America. 78 (6): 2025–2040.
21. http://internationalrivers.org/de/node/1477
22. Reservoir-Induced Seismicity
23. International Rivers
24. "A Human Trigger for the Great Quake of Sichuan?". Science. 323: 322. 2009. doi:10.1126/science.323.5912.322.
25. Chinese earthquake may have been man-made, say scientists, Telegraph, February 3, 2009
26. Naik, Gautam; Oster, Shai (February 6, 2009). "Scientists Link China's Dam to Earthquake, Renewing Debate". The Wall Street Journal.
27. Chen, L.; Talwani, P. (1998). "Seismicity in China". Pure and Applied Geophysics. 153: 133–149. doi:10.1007/s000240050188.
28. Redmayne, D.W. (1988) Mining induced seismicity in UK coalfields identified on the BGS National Seismograph Network. Geological Society, London, Engineering Geology Special Publications; vol. 5; pp. 405-413
29. http://bssa.geoscienceworld.org/cgi/content/abstract/101/1/327
30. Madrigal, Alexis (June 4, 2008). "Top 5 Ways to Cause a Man-Made Earthquake". Wired.
31. U.S. Geological Survey, Oklahoma - Magnitude 5.8.
32. Henry Fountain (March 28, 2013). "Study Links 2011 Quake to Technique at Oil Wells". The New York Times. Retrieved March 29, 2013.
33. Record tying Oklahoma earthquake felt as far away as Arizona, Associated Press, Ken Miller, September 3, 2016. Retrieved 3 September 2016.

34. USGS calls for shut down of wells, governor declares emergency in wake of 5.6 quake in Oklahoma, Enid News & Eagle, Sally Asher & Violet Hassler, September 3, 2016. Retrieved 4 September 2016.
35. Keller, G. Randy; Holland, Austin A. (March 22, 2013). Statement about the cause of 2011 Prague Earthquake Sequence (PDF). Oklahoma Geological Survey (Report). Retrieved April 30, 2015.
36. Pérez-Peña, Richard (April 23, 2015). "U.S. Maps Pinpoint Earthquakes Linked to Quest for Oil and Gas". New York Times. Retrieved November 8, 2015.
37. Andrews, Richard D.; Holland, Austin A. (April 21, 2015). Statement on Oklahoma Seismicity (PDF). Oklahoma Geological Survey (Report). University of Oklahoma. Retrieved April 30, 2015.
38. http://esd.lbl.gov/research/projects/induced_seismicity/oil&gas/
39. Van Eijsa, R.M.H.E; Muldersa, F.M.M; Nepveua, M; Kenterb, C.J; Scheffers, B.C. (2006). "Correlation between hydrocarbon reservoir properties and induced seismicity in the Netherlands". Engineering Geology. 84: 99–111. doi:10.1016/j.enggeo.2006.01.002.
40. González, P.J.; Tiampo K.F.; Palano M.; Cannavó F.; Fernández J. (2012). "The 2011 Lorca earthquake slip distribution controlled by groundwater crustal unloading". Nature Geoscience. 5: 821–825. doi:10.1038/ngeo1610. Retrieved 1 November 2012.
41. Tester, Jefferson W. (Massachusetts Institute of Technology); et al. (2006). The Future of Geothermal Energy - Impact of Enhanced Geothermal Systems (EGS) on the United States in the 21st Century (14MB PDF). Idaho Falls: Idaho National Laboratory. pp. 4–10. ISBN 0-615-13438-6. Retrieved 2007-02-07.
42. http://escholarship.org/uc/item/0t19709v The Impact of Injection on Seismicity at The Geysers Geothermal Field
43. Bromley, C.J. & Mongillo, M.A. (February 2007), "All Geothermal Energy from Fractured Reservoirs - Dealing with Induced Seismicity" (PDF), IEA Open Journal, 48 (7): 5
44. http://www.bssaonline.org/cgi/content/abstract/86/1A/93 Major earthquakes in Mexicali Valley, Mexico, and fluid extraction at Cerro Prieto Geothermal Field
45. Tester 2006, pp. 5–6
46. U.S. National Research Council Report, Induced Seismicity Potential in Energy Technologies, http://dels.nas.edu/Report/Induced-Seismicity-Potential-Energy-Technologies/13355
47. U.S. National Research Council Report-in-Brief, Induced Seismicity Potential in Energy Technologies, http://dels.nas.edu/Materials/Report-In-Brief/4298-Induced-Seismicity

CHAPTER 4

Measurement of Earthquakes

Chapter Contents

- **Introduction**
- **Seismic Scale**
- **Seismometer**
- **Earthquake Duration Magnitude**

The vibrations produced by earthquakes are detected, recorded, and measured by instruments call seismographs. The zig-zag line made by a seismograph, called a "seismogram," reflects the changing intensity of the vibrations by responding to the motion of the ground surface beneath the instrument. From the data expressed in seismograms, scientists can determine the time, the epicenter, the focal depth, and the type of faulting of an earthquake and can estimate how much energy was released.

INTRODUCTION

We can determine how big an earthquake is by measuring the size of the signal directly from the seismogram.

However, we also have to know how far away the earthquake was. This is because the amplitude of the seismic waves decreases with distance, so we must correct for this.

Today, an earthquake's size is typically reported simply by its magnitude, which is a measure of the size of the earthquake's source, where the ground began shaking.

While there are many modern scales used to calculate the magnitude, the most common is the moment magnitude, which allows for more precise measurements of large earthquakes than the Richter scale.

SEISMIC SCALE

Earthquakes are measured with two kinds of seismic scales: scales of the magnitude of the energy released by the rupture, and scales of the intensity of the resulting ground shaking at a given location.

Magnitude and Intensity

The severity of an earthquake is described by both magnitude and intensity. These two frequently confused terms refer to different, but related, expressions. Magnitude, usually expressed as an Arabic numeral characterizes the size of an earthquake by measuring indirectly the energy released. By contrast, intensity indicates the local effects and potential for damage produced by an earthquake on the Earth's surface as it affects humans, animals, structures, and natural objects such as bodies of water. Intensities are usually expressed in Roman numerals, and represent the severity of the shaking resulting from an earthquake. Ideally, any given earthquake can be described by only one magnitude, but many intensities since the earthquake effects vary with circumstances such as distance from the epicenter and local soil conditions. In practice, the same earthquake might have magnitude estimates typically differing by few tenths of a unit, depending on which magnitude scale is used and which data are included in the analysis.

Charles Richter, the creator of the Richter magnitude scale, distinguished intensity and magnitude as follows: "I like to use the analogy with radio transmissions. It applies in seismology because seismographs, or the receivers, record the waves of elastic disturbance, or radio waves, that are radiated from the earthquake source, or the broadcasting station. Magnitude can be compared to the power output in kilowatts of a broadcasting station. Local intensity on the Mercalli scale is then comparable to the signal strength on a receiver at a given locality; in effect, the quality of the signal. Intensity, like signal strength, will generally fall off with distance from the source, although it also depends on the local conditions and the pathway from the source to the point."

Seismic Intensity Scales

The first simple classification of earthquake intensity was devised by Domenico Pignataro in the 1780s. However, the first recognisable intensity scale in the modern sense of the word was drawn up by P.N.G. Egen in 1828; it was ahead of its time. The first widely adopted intensity scale, the Rossi–Forel scale, was introduced in the late 19th century. Since then numerous intensity scales have been developed and are used in different parts of the world.

Country/Region	Seismic intensity scale used
China	Liedu scale (GB/T 17742-1999)
Europe	European Macroseismic Scale (EMS-98)
Hong Kong	Modified Mercalli scale (MM)

India	Medvedev–Sponheuer–Karnik scale
Israel	Medvedev–Sponheuer–Karnik scale (MSK-64)
Japan	Shindo scale
Kazakhstan	Medvedev–Sponheuer–Karnik scale (MSK-64)
Russia	Medvedev–Sponheuer–Karnik scale (MSK-64)
Taiwan	Shindo scale
United States	Modified Mercalli scale (MM)

Unlike magnitude scales, intensity scales do not have a mathematical basis; instead they are an arbitrary ranking based on observed effects. Most seismic intensity scales have twelve degrees of intensity and are roughly equivalent to one another in values but vary in the degree of sophistication employed in their formulation.

Magnitude Scales

The first attempt to qualitatively define a single, absolute value to describe the size of earthquakes was the magnitude scale (the name being inspired by scales used to represent the brightness of stars).

Local Magnitude Scale and Related Scales

The local magnitude scale (M_L), also popularly known as the Richter scale, is a quantitative logarithmic scale. In the 1930s, California seismologist Charles F. Richter devised a simple numerical scale to describe the relative sizes of earthquakes in Southern California. The name "Richter scale" was coined by journalists and is not generally used by seismologists in technical literature. M_L is obtained by measuring the maximum displacement amplitude of a recording on a Wood–Anderson torsion seismometer (or one calibrated to it) at a distance of up to 600 km from the epicenter of the earthquake. Other more recent magnitude measurements include: body wave magnitude (m_b), surface wave magnitude (M_s), and duration magnitude (M_D). Each of these is scaled to give values similar to those given by the local magnitude scale; but because each is based on a measurement of one aspect of the seismogram, they do not always capture the overall power of the source. Specifically, some can be affected by saturation at higher magnitude values—meaning that they systematically underestimate the magnitude of larger events. This problem sets in at around magnitude 6 for local magnitude; surface-wave magnitude saturates above 8. Despite the limitations of older magnitude scales, they are still in wide use, as they can be calculated rapidly, catalogues of them dating back many years are available, they are sufficient for the vast majority of observed events, and the public is familiar with them.

Moment Magnitude Scale

Because of the limitations of the magnitude scales, a new, more uniformly applicable extension of them, known as moment magnitude (M_W) scale for representing the size of earthquakes, was introduced by Thomas C. Hanks and Hiroo Kanamori in 1977. In particular, for very large earthquakes moment magnitude gives the most reliable estimate of earthquake size. This is because seismic moment is derived from the concept of moment in physics and therefore provides clues to the physical size of an earthquake—the size of fault rupture and accompanying slip displacement—as well as the amount of energy released. So while seismic moment, too, is calculated from seismograms, it can also be obtained by working backwards from geologic estimates of the size of the fault rupture and displacement. The values of moments for observed earthquakes range over more than 15 orders of magnitude, and because they are not influenced by variables such as local circumstances, the results obtained make it easy to objectively compare the sizes of different earthquakes.

SEISMOMETER

Seismometers are instruments that measure motion of the ground, including those of seismic waves generated by earthquakes, volcanic eruptions, and other seismic sources. Records of seismic waves allow seismologists to map the interior of the Earth, and locate and measure the size of these different sources.

The word derives from the Greek, seismós, a shaking or quake, from the verb σείω, seíō, to shake; and μέτρον, métron, measure and was coined by David Milne-Home in 1841, to describe an instrument designed by Scottish physicist James David Forbes.

Japan Meteorological Agency Optical Electromagnetic Seismometer

Seismograph is another Greek term from seismós and γράφω, gráphō, to draw. It is often used to mean seismometer, though it is more applicable to the older instruments in which the measuring and recording of ground motion were combined than to modern systems, in which these functions are separated. Both types provide a continuous record of ground motion; this distinguishes them from seismoscopes, which merely indicate that motion has occurred, perhaps with some simple measure of how large it was.

The concerning technical discipline is called seismometry, a branch of seismology.

Basic Principles

A simple seismometer that is sensitive to up-down motions of the earth can be understood by visualizing a weight hanging on a spring. The spring and weight are suspended from a frame that moves along with the earth's surface. As the earth moves, the relative motion between the weight and the earth provides a measure of the vertical ground motion. If a recording system is installed, such as a rotating drum attached to the frame, and a pen attached to the mass, this relative motion between the weight and earth can be recorded to produce a history of ground motion, called a seismogram.

Any movement of the ground moves the frame. The mass tends not to move because of its inertia, and by measuring the movement between the frame and the mass, the motion of the ground can be determined.

Early seismometers used optical levers or mechanical linkages to amplify the small motions involved, recording on soot-covered paper or photographic paper. Modern instruments use electronics. In some systems, the mass is held nearly motionless relative to the frame by an electronic negative feedback loop. The motion of the mass relative to the frame is measured, and the feedback loop applies a magnetic or electrostatic force to keep the mass nearly motionless. The voltage needed to produce this force is the output of the seismometer, which is recorded digitally. In other systems the weight is allowed to move, and its motion produces a voltage in a coil attached to the mass and moving through the magnetic field of a magnet attached to the frame. This design is often used in the geophones used in seismic surveys for oil and gas.

Professional seismic observatories usually have instruments measuring three axes: north-south (y-axis), east-west (x-axis), and the vertical (z-axis). If only one axis is measured, this is usually the vertical because it is less noisy and gives better records of some seismic waves.

The foundation of a seismic station is critical. A professional station is sometimes mounted on bedrock. The best mountings may be in deep boreholes, which avoid thermal effects, ground noise and tilting from weather and tides. Other instruments are often mounted in insulated enclosures on small buried piers of unreinforced concrete. Reinforcing rods and aggregates would distort the pier as the temperature changes. A site is always surveyed for ground noise with a temporary installation before pouring the pier and laying conduit. Originally, European seismographs were placed in a particular area after a destructive earthquake. Today, they are spread to provide appropriate coverage (in the case of weak-motion seismology) or concentrated in high-risk regions (strong-motion seismology).

History

Ancient Era

In AD 132, Zhang Heng of China's Han dynasty invented the first seismoscope (by the definition above), which was called Houfeng Didong Yi (translated as, "instrument for measuring the seasonal winds and the movements of the Earth"). The description we have, from the History of the Later Han Dynasty, says that it was a large bronze vessel, about 2 meters in diameter; at eight points around the top were dragon's heads holding bronze balls. When there was an earthquake, one of the mouths would open and drop its ball into a bronze toad at the base, making a sound and supposedly showing the direction of the earthquake.

Replica of Zhang Heng's seismoscope Houfeng Didong Yi

On at least one occasion, probably at the time of a large earthquake in Gansu in AD 143, the seismoscope indicated an earthquake even though one was not felt. The available text says that inside the vessel was a central column that could move along eight tracks; this is thought to refer to a pendulum, though it is not known exactly how this was linked to a mechanism that would open only one dragon's mouth. The first ever earthquake recorded by this seismoscope was supposedly somewhere in the east. Days later, a rider from the east reported this earthquake.

Modern Designs

The principle can be shown by an early special purpose seismometer. This consisted of a large stationary pendulum, with a stylus on the bottom. As the earth starts to move, the heavy mass of the pendulum has the inertia to stay still in the non-earth frame of reference. The result is that the stylus scratches a pattern corresponding with the Earth's movement. This type of strong motion seismometer recorded upon a smoked glass (glass with carbon soot). While not sensitive enough to detect distant earthquakes, this instrument could indicate the direction of the pressure waves and thus help find the epicenter of a local earthquake – such instruments were useful in the analysis of the 1906 San Francisco earthquake. Further re-analysis was performed in the 1980s using these early recordings, enabling a more precise determination of the initial fault break location in Marin county and its subsequent progression, mostly to the south.

Milne horizontal pendulum seismometer. One of the Important Cultural Properties of Japan. Exhibit in the National Museum of Nature and Science, Tokyo, Japan.

After 1880, most seismometers were descended from those developed by the team of John Milne, James Alfred Ewing and Thomas Gray, who worked in Japan from 1880 to 1895. These seismometers used damped horizontal pendulums. After World War II, these were adapted into the widely used Press-Ewing seismometer.

Later, professional suites of instruments for the worldwide standard seismographic network had one set of instruments tuned to oscillate at fifteen seconds, and the other at ninety seconds, each set measuring in three directions. Amateurs or observatories with limited means tuned their smaller, less sensitive instruments to ten seconds. The basic damped horizontal pendulum seismometer swings like the gate of a fence. A heavy weight is mounted on the point of a long (from 10 cm to several meters) triangle, hinged at its vertical edge. As the ground moves, the weight stays unmoving, swinging the "gate" on the hinge.

The advantage of a horizontal pendulum is that it achieves very low frequencies of oscillation in a compact instrument. The "gate" is slightly tilted, so the weight tends to slowly return to a central position. The pendulum is adjusted (before the damping is installed) to oscillate once per three seconds, or once per thirty seconds. The general-purpose instruments of small stations or amateurs usually oscillate once per ten seconds. A pan of oil is placed under the arm, and a small sheet of metal mounted on the underside of the arm drags in the oil to damp oscillations. The level of oil, position on the arm, and angle and size of sheet is adjusted until the damping is "critical," that is, almost having oscillation. The hinge is very low friction, often torsion wires, so the only friction is the internal friction of the wire. Small seismographs with low proof masses are placed in a vacuum to reduce disturbances from air currents. Zollner described torsionally suspended horizontal pendulums as early as 1869, but developed them for gravimetry rather than seismometry. Early seismometers had an arrangement of levers on jeweled bearings, to scratch smoked glass or paper. Later, mirrors reflected a light beam to a direct-recording plate or roll of photographic paper. Briefly, some designs returned to mechanical movements to save money. In mid-twentieth-century systems, the light was reflected to a pair of differential electronic photosensors called a photomultiplier. The voltage generated in the photomultiplier was used to drive galvanometers which had a small mirror mounted on the axis. The moving reflected light beam would strike the surface of the turning drum, which was covered with photo-sensitive paper. The expense of developing photo sensitive paper caused many seismic observatories to switch to ink or thermal-sensitive paper.

Modern Instruments

Modern instruments use electronic sensors, amplifiers, and recording devices. Most are broadband covering a wide range of frequencies. Some seismometers can measure motions with frequencies from 500 Hz to 0.00118 Hz (1/500 = 0.002 seconds per cycle, to 1/0.00118 = 850 seconds per cycle). The mechanical suspension for horizontal instruments remains the garden-gate described above. Vertical instruments use some kind of constant-force suspension, such as the LaCoste suspension. The LaCoste suspension uses a zero-length spring to provide a long period (high sensitivity). Some modern instruments use a "triaxial" design, in which three identical motion sensors are set at the same angle to the vertical but 120 degrees apart on the horizontal. Vertical and horizontal motions can be computed from the outputs of the three sensors.

CMG-40T triaxial broadband seismometer

Seismometers unavoidably introduce some distortion into the signals they measure, but professionally designed systems have carefully characterized frequency transforms.

Modern sensitivities come in three broad ranges: geophones, 50 to 750 V/m; local geologic seismographs, about 1,500 V/m; and teleseismographs, used for world survey, about 20,000 V/m. Instruments come in three main varieties: short period, long period and broadband. The short and long period measure velocity and are very sensitive, however they 'clip' the signal or go off-scale for ground motion that is strong enough to be felt by people. A 24-bit analog-to-digital conversion channel is commonplace. Practical devices are linear to roughly one part per million.

Delivered seismometers come with two styles of output: analog and digital. Analog seismographs require analog recording equipment, possibly including an analog-to-digital converter. The output of a digital seismograph can be simply input to a computer. It presents the data in a standard digital format (often "SE2" over Ethernet).

Teleseismometers

The modern broadband seismograph can record a very broad range of frequencies. It consists of a small "proof mass", confined by electrical forces, driven by sophisticated electronics. As the earth moves, the electronics attempt to hold the mass steady through a feedback circuit. The amount of force necessary to achieve this is then recorded.

In most designs the electronics holds a mass motionless relative to the frame. This device is called a "force balance accelerometer". It measures acceleration instead of velocity of ground movement. Basically, the distance between the mass and some part of the frame is measured very precisely, by a linear variable differential transformer. Some instruments use a linear variable differential capacitor.

A low-frequency 3-direction ocean-bottom seismometer (cover removed). Two masses for x- and y-direction can be seen, the third one for z-direction is below. This model is a CMG-40TOBS, manufactured by Güralp Systems Ltd and is part of the Monterey Accelerated Research System.

That measurement is then amplified by electronic amplifiers attached to parts of an electronic negative feedback loop. One of the amplified currents from the negative feedback loop drives a coil very like a loudspeaker, except that the coil is attached to the mass, and the magnet is mounted on the frame. The result is that the mass stays nearly motionless.

Most instruments measure directly the ground motion using the distance sensor. The voltage generated in a sense coil on the mass by the magnet directly measures the instantaneous velocity of the ground. The current to the drive coil provides a sensitive, accurate measurement of the force between the mass and frame, thus measuring directly the ground's acceleration (using f=ma where f=force, m=mass, a=acceleration).

One of the continuing problems with sensitive vertical seismographs is the buoyancy of their masses. The uneven changes in pressure caused by wind blowing on an open window can easily change the density of the air in a room enough to cause a vertical seismograph to show spurious signals. Therefore, most professional seismographs are sealed in rigid gas-tight enclosures. For example, this is why a common Streckeisen model has a thick glass base that must be glued to its pier without bubbles in the glue.

It might seem logical to make the heavy magnet serve as a mass, but that subjects the seismograph to errors when the Earth's magnetic field moves. This is also why seismograph's moving parts are constructed from a material that interacts minimally with magnetic fields. A seismograph is also sensitive to changes in temperature so many instruments are constructed from low expansion materials such as nonmagnetic invar.

The hinges on a seismograph are usually patented, and by the time the patent has expired, the design has been improved. The most successful public domain designs use thin foil hinges in a clamp.

Another issue is that the transfer function of a seismograph must be accurately characterized, so that its frequency response is known. This is often the crucial difference between professional and amateur instruments. Most instruments are characterized on a variable frequency shaking table.

Strong-Motion Seismometers

Another type of seismometer is a digital strong-motion seismometer, or accelerograph. The data from such an instrument is essential to understand how an earthquake affects manmade structures.

A strong-motion seismometer measures acceleration. This can be mathematically integrated later to give velocity and position. Strong-motion seismometers are not as sensitive to ground motions as teleseismic instruments but they stay on scale during the strongest seismic shaking.

Other Forms

Accelerographs and geophones are often heavy cylindrical magnets with a spring-mounted coil inside. As case moves, the coil tends to stay stationary, so the magnetic field cuts the wires, inducing current in the output wires. They receive frequencies from several hundred hertz down to 1 Hz. Some have electronic damping, a low-budget way to get some of the performance of the closed-loop wide-band geologic seismographs.

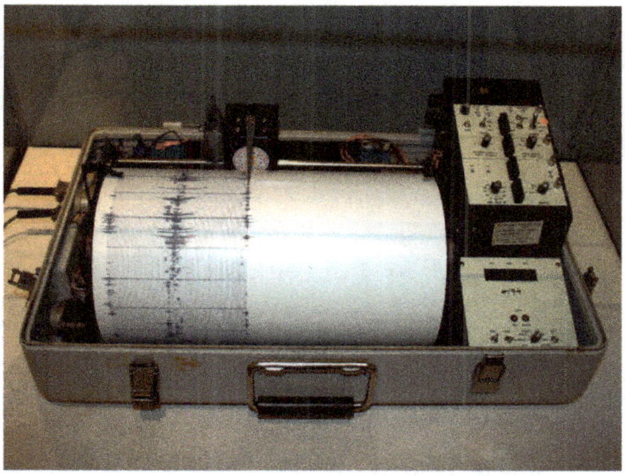

A Kinemetrics seismograph, formerly used by the United States Department of the Interior.

Strain-beam accelerometers constructed as integrated circuits are too insensitive for geologic seismographs (2002), but are widely used in geophones.

Some other sensitive designs measure the current generated by the flow of a non-corrosive ionic fluid through an electret sponge or a conductive fluid through a magnetic field.

Interconnected Seismometers

Seismometers spaced in an array can also be used to precisely locate, in three dimensions, the source of an earthquake, using the time it takes for seismic waves to propagate away from the hypocenter, the initiating point of fault rupture. Interconnected seismometers are also used to detect underground nuclear test explosions, as well as for Earthquake early warning systems. These seismometer are often used as part of a large scale governmental or scientific project, but some organizations such as the Quake-Catcher Network, can use residential size detectors built into computers to detect earthquakes as well.

In reflection seismology, an array of seismometers image sub-surface features. The data are reduced to images using algorithms similar to tomography. The data reduction methods resemble those of computer-aided tomographic medical imaging X-ray machines (CAT-scans), or imaging sonars.

A worldwide array of seismometers can actually image the interior of the Earth in wave-speed and transmissivity. This type of system uses events such as earthquakes, impact events or nuclear explosions as wave sources. The first efforts at this method used manual data reduction from paper seismograph charts. Modern digital seismograph records are better adapted to direct computer use. With inexpensive seismometer designs and internet access, amateurs and small institutions have even formed a "public seismograph network."

Seismographic systems used for petroleum or other mineral exploration historically used an explosive and a wireline of geophones unrolled behind a truck. Now most short-range systems use "thumpers" that hit the ground, and some small commercial systems have such good digital signal processing that a few sledgehammer strikes provide enough signal for short-distance refractive surveys. Exotic cross or two-dimensional arrays of geophones are sometimes used to perform three-dimensional reflective imaging of subsurface features. Basic linear refractive geomapping software (once a black art) is available off-the-shelf, running on laptop computers, using strings as small as three geophones. Some systems now come in an 18” (0.5 m) plastic field case with a computer, display and printer in the cover.

Small seismic imaging systems are now sufficiently inexpensive to be used by civil engineers to survey foundation sites, locate bedrock, and find subsurface water.

Recording

Today, the most common recorder is a computer with an analog-to-digital converter, a disk drive and an internet connection; for amateurs, a PC with a sound card and associated software is adequate. Most systems record continuously, but some record only when a signal is detected, as shown by a short-term increase in the variation of the signal, compared to its long-term average (which can vary slowly because of changes in seismic noise).

Viewing of a Develocorder film

Matsushiro Seismological Observatory

Prior to the availability of digital processing of seismic data in the late 1970s, the records were done in a few different forms on different types of media. A "Helicorder" drum was a device used to record data into photographic paper or in the form of paper and ink. A "Develocorder" was a machine that record data from up to 20 channels into a 16-mm film. The recorded film can be viewed by a machine. The reading and measuring from these types of media can be done by hand. After the digital processing has been used, the archives of the seismic data were recorded in magnetic tapes. Due to the deterioration of older magnetic tape medias, large number of waveforms from the archives are not recoverable.

EARTHQUAKE DURATION MAGNITUDE

The concept of Earthquake Duration Magnitude - originally proposed by Bisztricsany in 1958 using surface waves only - is based on the realization that on a recorded earthquake seismogram, the total length of the seismic wavetrain - sometimes referred to as the CODA - reflects its size. Thus larger earthquakes give longer seismograms [as well as stronger seismic waves] than small ones. The seismic wave interval measured on the time axis of an earthquake record - starting with the first seismic wave onset until the wavetrain amplitude diminishes to at least 10% of its maximum recorded value - is referred to as "earthquake duration". It is this concept that Bisztricsany first used to develop his Earthquake Duration Magnitude Scale employing surface wave durations.

Earthquake Duration Magnitude [Md] Development

In 1965, Solovev proposed the use of total duration instead of the duration of surface waves. In 1972, Lee et al. used coda duration for the first time to estimate Richter magnitude of local Californian earthquakes. Based on their study, they suggested that it is appropriate to estimate the magnitude of local earthquakes using signal duration. More recently, the development in instrumentation led to the use of signal duration to estimate the coda magnitude (Md) for earthquakes recorded on short-period vertical seismographs. Numerous studies determined the relation between coda duration and magnitude for different regions of the World. According to a recent study by Mandal et al. (2004), previous studies showed that duration magnitude estimation is quite stable for local earthquakes ranging from magnitude Md 0.0 to 5.0.

Md Empirical Relationships

In two most recent investigations using statistically stable samples for Italian earthquakes [approximately 100,000 events over the period 1981-2002 in the Richter local [ML] magnitude range of 3.5-5.8] and for Indian earthquakes exemplified by an aftershock sequence of 121 events with Ms (surface wave magnitude) > 4.0 in 2001 in the Bhuj area of northwestern India, the latest empirically derived equations for Md determinations are published:

$$Md = 2.49log10(T) - 2.31 + StationCorrectionFactor$$ (Castello et al., 2007)

$$Md = 0.80log10(T)^2 + 1.7log10(T) - 0.87$$ (Mandal et al., 2004)

ML from Md

Although conversions between empirically derived "sensitive" seismic parameters such as earthquake magnitude scales is mathematically cautioned as well as physically limited, some seismologists such as Brumbaugh have nevertheless attempted to produce a relationship linking ML to Md for a rather small sample of 17 events in Arizona:

$$ML = 0.936Md - 0.16 + / - 0.22$$

REFERENCES

1. Spall H. "Charles F. Richter - An Interview". USGS. Retrieved 4 December 2014.
2. David Alexander (1993). Natural Disasters (First ed.). Springer Science+Business Media. p. 28. ISBN 978-0-412-04741-1.
3. "The European Macroseismic Scale EMS-98". Centre Européen de Géodynamique et de Séismologie (ECGS). Retrieved 2013-07-26.
4. "Magnitude and Intensity of an Earthquake". Hong Kong Observatory. Retrieved 2008-09-15.
5. "The Severity of an Earthquake". U.S. Geological Survey. Retrieved 2012-01-15.
6. Hough, S.E. (2007). Richter's scale: measure of an earthquake, measure of a man. Princeton University Press. p. 121. ISBN 978-0-691-12807-8. Retrieved 10 December 2011.
7. Shearer, P.M., Introduction to Seismology, Cambridge University Press, Cambridge, 1999.
8. Ben-Menahem, A. (2009). Historical Encyclopedia of Natural and Mathematical Sciences , Volume 1. Springer. p. 2657. ISBN 9783540688310. Retrieved 28 August 2012.
9. Richter, C.F. (1958). Elementary Seismology. San Francisco: W.H. Freeman.
10. William H.K. Lee; Paul Jennings; Carl Kisslinger; Hiroo Kanamori (27 September 2002). International Handbook of Earthquake & Engineering Seismology. Academic Press. pp. 283–. ISBN 978-0-08-048922-3. Retrieved 29 April 2013.
11. Erhard Wielandt's 'Seismic Sensors and their Calibration'- Current (2002) reference by a widely consulted expert.
12. Reitherman, Robert (2012). Earthquakes and Engineers: an International History. Reston, VA: ASCE Press. pp. 122–125. ISBN 9780784410714.
13. Sleeswyk AW, Sivin N (1983). "Dragons and toads: the Chinese seismoscope of BC. 132". Chinese Science. 6: 1–19.

14. Needham, Joseph (1959). Science and Civilization in China, Volume 3: Mathematics and the Sciences of the Heavens and the Earth. Cambridge: Cambridge University Press. pp. 626–635.
15. Zero Length Springs in Seismographs
16. A Biography of Lucien LaCoste, inventor of the zero-length spring
17. Public seismograph network- many resources for amateurs and underfunded institutions
18. Hutton, Kate; Yu, Ellen. "NEWS FLASH!! SCSN Earthquake Catalog Completed!!" (PDF). Seismological Laboratory, Caltech. Retrieved 4 July 2014.
19. Fogleman, Kent A.; Lahr, John C.; Stephens, Christopher D.; Page, Robert A. (June 1993). Earthquake Locations Determined by the Southern Alaska Seismograph Network for October 1971 through May 1989 (Report). USGS.
20. Bisztricsany, E. A., 1958, A new method for the determination of the magnitude of earthquakes. Geofiz. Kozl., 69-76.
21. Solov'ev, S. L., Seismicity of Sakalin. Bull. Earth Res. Inst.,Tokyo Univ., 1965, 43, 95–102.
22. Lee, W. H. K., Bennet, R. E. and Meaghu, K. L., A method of estimating magnitude of local earthquakes from signal duration. U.S. Geological Survey Open File Report, 1972, 28 pp.
23. Mandal, P., Narasaiah, R., Raju P. S., & R. K. Chandha, 2004. Coda duration magnitude scale of 2001 Bhuj aftershoks, India. Current Science, Vol. 87, No. 4, 2004. (India).
24. Castello, B., Olivieri, M. & G. Selvaggi, 2007. Local and duration magnitude determination for the Italian earthquake catalog, 1981-2002.
25. Mandal et al. 2004 (op. cit.)
26. Brumbaugh, D. S., 1989. A comparison of duration magnitude to local magnitude for seismic events recorded in northern Arizona. Jour. Arizona-Nevada Acad. Sci., Vol. 23, 29-31

CHAPTER 5

Seismic Base Isolation

Chapter Contents

Base isolation is a state-of-the-art method in which the structure (superstructure) is separated from the base (foundation or substructure) by introducing a suspension system between the base and the main structure.

In context of seismic design of structures, base isolation can be replaced with seismic isolation i.e., the structure above the ground, which is most affected during earthquake is separated from the effects of earthquake forces by introducing a mechanism that will help the structure to hover. The concept of base isolation is quite easy to grasp. It can be explained as a bird flying during an earthquake is not affected. In simple words if structure is floating on its base, the movement of ground will have no effect on the structure.

INTRODUCTION

Base isolation, also known as seismic base isolation or base isolation system, is one of the most popular means of protecting a structure against earthquake forces. It is a collection of structural elements which should substantially decouple a superstructure from its substructure resting on a shaking ground thus protecting a building or non-building structure's integrity.

Base isolation is one of the most powerful tools of earthquake engineering pertaining to the passive structural vibration control technologies. It is meant to enable a building or non-building structure to survive a potentially devastating seismic impact through a proper initial design or subsequent modifications. In some cases, application of base isolation can raise both a structure's seismic performance and its seismicsustainability considerably. Contrary to popular belief base isolation does not make a building earthquake proof.

Base isolation system consists of *isolation units* with or without *isolation components*, where:

- *Isolation units* are the basic elements of a *base isolation system* which are intended to provide the aforementioned decoupling effect to a building or non-building structure.
- *Isolation components* are the connections between *isolation units* and their parts having no decoupling effect of their own.

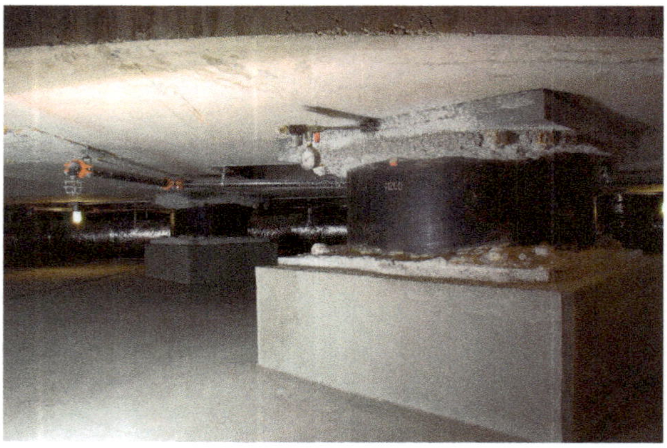

The base isolators under the Utah State Capitol building

LA City Hall, the tallest base isolated building in the world

Isolation units could consist of shear or sliding units.

This technology can be used for both new structural design and seismic retrofit. In process of seismic retrofit, some of the most prominent U.S. monuments, e.g. Pasadena City Hall, San Francisco City Hall, Salt Lake City and County Building or LA City Hall were mounted on *base isolation systems*. It required creating rigidity diaphragms and moats around the buildings, as well as making provisions against overturning and P-Delta Effect.

Base isolation is also used on a smaller scale—sometimes down to a single room in a building. Isolated raised-floor systems are used to safeguard essential equipment against earthquakes. The technique has been incorporated to protect statues and other works of art—see, for instance, Rodin's *Gates of Hell* at the National Museum of Western Art in Tokyo's Ueno Park.

Base isolation demonstration at The Field Museum in Chicago

HISTORY

The first evidence of architects using the principle of base isolation for earthquake protection was discovered in Pasargadae, a city in ancient Persia, now Iran: it goes back to 6th century BC.

Tomb of Cyrus is said to be the oldest base-isolated structure in the world.

It works by having a wide and deep stone and mortar foundation, smoothed at the top, upon which a second foundation is built of wide, smoothed stones which are linked together, forming a plate that slides back and forth over the lower foundation in case of an earthquake, leaving the structure intact.

RESEARCH ON BASE ISOLATION

Through the George E. Brown, Jr. Network for Earthquake Engineering Simulation (NEES), researchers are studying the performance of base isolation systems. The project, a collaboration among researchers at University of Nevada, Reno; University of California, Berkeley; University of Wisconsin, Green Bay; and the University at Buffalo is conducting a strategic assessment of the economic, technical, and procedural barriers to the widespread adoption of seismic isolation in the United States. NEES resources have been used for experimental and numerical simulation, data mining, networking and collaboration to understand the complex interrelationship among the factors controlling the overall performance of an isolated structural system. This project involves earthquake shaking table and hybrid tests at the NEES experimental facilities at the University of California, Berkeley, and the University at Buffalo, aimed at understanding ultimate performance limits to examine the propagation of local isolation failures (e.g., bumping against stops, bearing failures, uplift) to the system level response. These tests, including a full-scale, three-dimensional test of an isolated 5-story steel building on the E-Defense shake table in Miki, Hyogo, Japan, will help fill critical knowledge gaps, validate assumptions regarding behavior and modeling, and provide essential proof-of-concept evidence regarding the importance of isolation technology.

Adaptive Base Isolation

An adaptive base isolation system includes a tunable isolator that can adjust its properties based on the input to minimize the transferred vibration. Magnetorheological fluid dampers and isolators with Magnetorheological elastomer have been suggested as adaptive base isolators.

REFERENCES

1. "Los Angeles City Hall Seismic Rehabilitation Project – Base Isolation Technology". Archived from the original on 27 July 2011.
2. Pressman, Andy (2007). Architectural graphic standards. John Wiley and Sons. p. 30. ISBN 978-0-471-70091-3.
3. Webster, Anthony C. (1994). Technological advance in Japanese building design and construction. ASCE Publications. p. 70. ISBN 978-0-87262-932-5.
4. Datta, T. K. (2010). Seismic Analysis of Structures. John Wiley and Sons. p. 369. ISBN 978-0-470-82462-7.
5. https://www.youtube.com/watch?v=ZqlXp3czrrM
6. Lead Rubber Bearing being tested at the UCSD Caltrans-SRMD facility, YouTube
7. Hybrid Simulation of Base Isolated Structures, YouTube
8. Seismic Isolation Projects in Japan
9. Reitherman, Robert (2012). Earthquakes and Engineers: An International History. Reston, VA: ASCE Press. ISBN 9780784410622.
10. Seismic isolation in buildings to be a practical reality: Behavior of structure and installation technique
11. nees@berkeley project highlight: NEES TIPS Seismic Isolation Hybrid Simulation, https://www.youtube.com/watch?v=Uh6l5Jqtp0c
12. Giovannardi, Fausto; Guisasola, Adriana (2013). "Base isolation: dalle origini ai giorni nostri". Retrieved October 7, 2013.
13. Yang, G., et al. "Large-scale MR fluid dampers: modeling and dynamic performance considerations." Engineering structures 24.3 (2002): 309-323. [link]
14. Behrooz, Majid, Xiaojie Wang, and Faramarz Gordaninejad. "Performance of a new magnetorheological elastomer isolation system." Smart Materials and Structures 23.4 (2014) [link]

CHAPTER 6

Vibration Isolation

Chapter Contents

Vibration can be defined as motion generated by any mass or force acting on that mass. In some instances, vibration can be the main cause of serious damages that can destroy a machine or can lead to a human's discomfort. Because of these issues, there are always many ways to provide vibration isolation, which means to lessen the impact of the vibration or to provide vibration control so that damage is minimized.

INTRODUCTION

Vibration isolation is the process of isolating an object, such as a piece of equipment, from the source of vibrations.

Vibration is undesirable in many domains, primarily engineered systems and habitable spaces, and methods have been developed to prevent the transfer of vibration to such systems. Vibrations propagate via mechanical waves and certain mechanical linkages conduct vibrations more efficiently than others. Passive vibration isolation makes use of materials and mechanical linkages that absorb and damp these mechanical waves. Active vibration isolation involves sensors and actuators that produce destructive interference that cancels-out incoming vibration.

PASSIVE ISOLATION

"Passive vibration isolation" refers to vibration isolation or mitigation of vibrations by passive techniques such as rubber pads or mechanical springs, as opposed to "active vibration isolation" or "electronic force cancellation" employing electric power, sensors, actuators, and control systems.

Passive vibration isolation is a vast subject, since there are many types of passive vibration isolators used for many different applications. A few of these applications are for industrial equipment such as pumps, motors, HVAC systems, or washing machines; isolation of civil engineering structures from earthquakes (base isolation), sensitive laboratory equipment, valuable statuary, and high-end audio.

A basic understanding of how passive isolation works, the more common types of passive isolators, and the main factors that influence the selection of passive isolators:

Common Passive Isolation Systems

Pneumatic or Air Isolators

These are bladders or canisters of compressed air. A source of compressed air is required to maintain them. Air springs are rubber bladders which provide damping as well as isolation and are used in large trucks. Some pneumatic isolators can attain low resonant frequencies and are used for isolating large industrial equipment. Air tables consist of a working surface or optical surface mounted on air legs. These tables provide enough isolation for laboratory instrument under some conditions. Air systems may leak under vacuum conditions. The air container can interfere with isolation of low-amplitude vibration.

Mechanical Springs and Spring-Dampers

These are heavy-duty isolators used for building systems and industry. Sometimes they serve as mounts for a concrete block, which provides further isolation.

Pads or Sheets of Flexible Materials Such As Elastomers, Rubber, Cork, Dense Foam and Laminate Materials.

Elastomer pads, dense closed cell foams and laminate materials are often used under heavy machinery, under common household items, in vehicles and even under higher performing audio systems.

Molded and Bonded Rubber and Elastomeric Isolators and Mounts

These are often used as machinery mounts or in vehicles. They absorb shock and attenuate some vibration.

Negative-Stiffness Isolators

Negative-stiffness isolators are less common than other types and have generally been developed for high-level research applications such as gravity wave detection. Lee, Goverdovskiy, and Temnikov (2007) proposed a negative-stiffness system for isolating vehicle seats.

The focus on negative-stiffness isolators has been on developing systems with very low resonant frequencies (below 1 Hz), so that low frequencies can be adequately isolated, which is critical for sensitive instrumentation. All higher frequencies are also isolated. Negative-stiffness systems can be made with low stiction, so that they are effective in isolating low-amplitude vibrations. Negative-stiffness mechanisms are purely mechanical and typically involve the configuration and loading of components such as beams or inverted pendulums. Greater loading of the negative-stiffness mechanism, within the range of its operability, decreases the natural frequency.

Wire Rope Isolators

These isolators are durable and can withstand extreme environments. They are often used in military applications.

Base Isolators for Seismic Isolation of Buildings, Bridges, etc.

Base isolators made of layers of neoprene and steel with a low horizontal stiffness are used to lower the natural frequency of the building. Some other base isolators are designed to slide, preventing the transfer of energy from the ground to the building.

Tuned Mass Dampers

Tuned mass dampers reduce the effects of harmonic vibration in buildings or other structures. A relatively small mass is attached in such a way that it can dampen out a very narrow band of vibration of the structure.

Do it Yourself Isolators

In less sophisticated solutions, bungee cords can be used as a cheap isolation system which may be effective enough for some applications. The item to be isolated is suspended from the bungee cords. This is difficult to implement without a danger of the isolated item falling. Tennis balls cut in half have been used under washing machines and other items with some success.

How Passive Isolation Works

A passive isolation system, such as a shock mount, in general contains mass, spring, and damping elements and moves as a harmonic oscillator. The mass and spring stiffness dictate a natural frequency of the system. Damping causes energy dissipation and has a secondary effect on natural frequency.

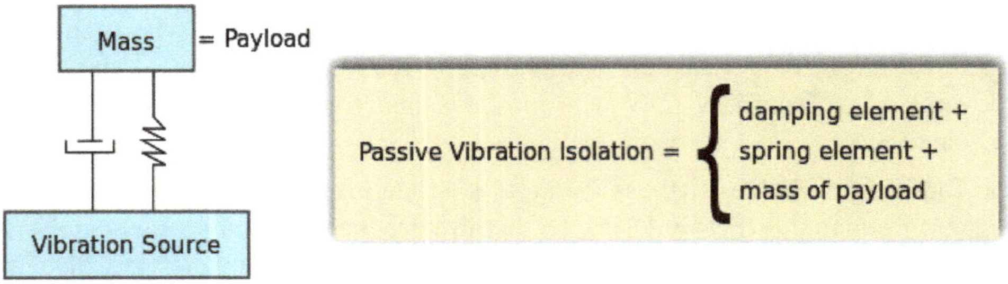

Passive Vibration Isolation

Every object on a flexible support has a fundamental natural frequency. When vibration is applied, energy is transferred most efficiently at the natural frequency, somewhat efficiently below the natural frequency, and with increasing inefficiency (decreasing efficiency) above the natural frequency. This can be seen in the transmissibility curve, which is a plot of transmissibility vs. frequency.

Here is an example of a transmissibility curve. Transmissibility is the ratio of vibration of the isolated surface to that of the source. Vibrations are never completely eliminated, but they can be greatly reduced. The curve below shows the typical performance of a passive, negative-stiffness isolation system with a natural frequency of 0.5 Hz. The general shape of the curve is typical for passive systems. Below the natural frequency, transmissibility hovers near 1. A value of 1 means that vibration is going through the system without being amplified or reduced. At the resonant frequency, energy is transmitted efficiently, and the incoming vibration is amplified. Damping in the system limits the level of amplification. Above the resonant frequency, little energy can be transmitted, and the curve rolls off to a low value. A passive isolator can be seen as a mechanical low-pass filter for vibrations.

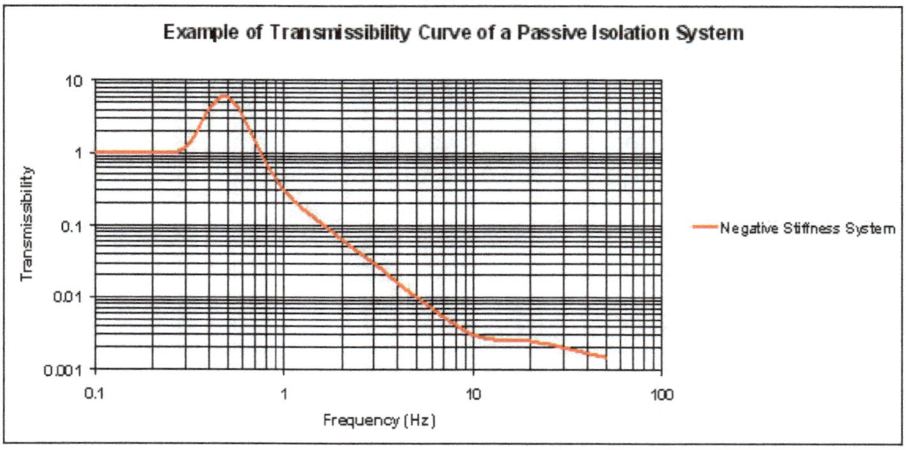

Negative-stiffness transmissibility

In general, for any given frequency above the natural frequency, an isolator with a lower natural frequency will show greater isolation than one with a higher natural frequency. The best isolation system for a given situation depends on the frequency, direction, and magnitude of vibrations present and the desired level of attenuation of those frequencies.

All mechanical systems in the real world contain some amount of damping. Damping dissipates energy in the system, which reduces the vibration level which is transmitted at the natural frequency. The fluid in automotive shock absorbers is a kind of damper, as is the inherent damping in elastomeric (rubber) engine mounts.

Damping is used in passive isolators to reduce the amount of amplification at the natural frequency. However, increasing damping tends to reduce isolation at the higher frequencies. As damping is increased, transmissibility roll-off decreases. This can be seen in the chart below.

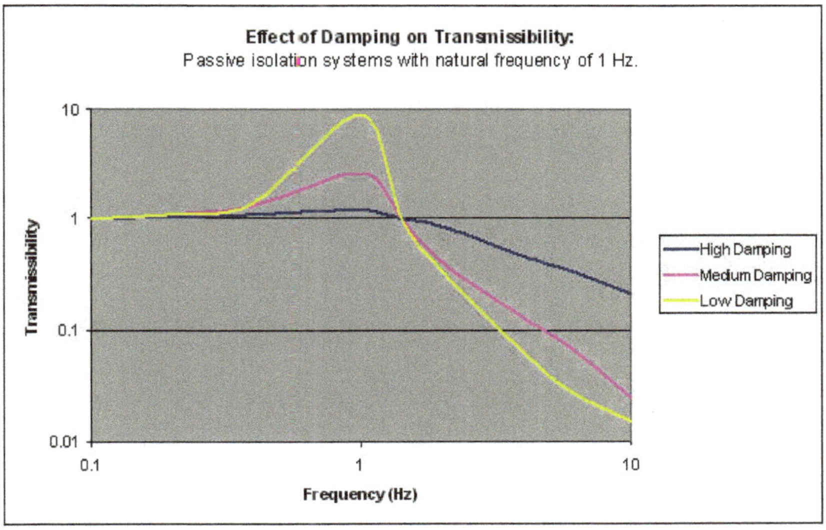

Damping effect on transmissibility

Passive isolation operates in both directions, isolating the payload from vibrations originating in the support, and also isolating the support from vibrations originating in the payload. Large machines such as washers, pumps, and generators, which would cause vibrations in the building or room, are often isolated from the floor. However, there are a multitude of sources of vibration in buildings, and it is often not possible to isolate each source. In many cases, it is most efficient to isolate each sensitive instrument from the floor. Sometimes it is necessary to implement both approaches.

Factors Influencing the Selection of Passive Vibration Isolators

- Characteristics of item to be isolated
  - Size: The dimensions of the item to be isolated help determine the type of isolation which is available and appropriate. Small objects may use only one

isolator, while larger items might use a multiple-isolator system.

- Weight: The weight of the object to be isolated is an important factor in choosing the correct passive isolation product. Individual passive isolators are designed to be used with a specific range of loading.

- Movement: Machines or instruments with moving parts may affect isolation systems. It is important to know the mass, speed, and distance traveled of the moving parts.

■ Operating Environment

- Industrial: This generally entails strong vibrations over a wide band of frequencies and some amount of dust.

- Laboratory: Labs are sometimes troubled by specific building vibrations from adjacent machinery, foot traffic, or HVAC airflow.

- Indoor or outdoor: Isolators are generally designed for one environment or the other.

- Corrosive/non-corrosive: Some indoor environments may present a corrosive danger to isolator components due to the presence of corrosive chemicals. Outdoors, water and salt environments need to be considered.

- Clean room: Some isolators can be made appropriate for clean room.

- Temperature: In general, isolators are designed to be used in the range of temperatures normal for human environments. If a larger range of temperatures is required, the isolator design may need to be modified.

- Vacuum: Some isolators can be used in a vacuum environment. Air isolators may have leakage problems. Vacuum requirements typically include some level of clean room requirement and may also have a large temperature range.

- Magnetism: Some experimentation which requires vibration isolation also requires a low-magnetism environment. Some isolators can be designed with low-magnetism components.

- Acoustic noise: Some instruments are sensitive to acoustic vibration. In addition, some isolation systems can be excited by acoustic noise. It may be necessary to use an acoustic shield. Air compressors can create problematic acoustic noise, heat, and airflow.

- Static or dynamic loads: This distinction is quite important as isolators are designed for a certain type and level of loading.

Static Loading

Static Loading is basically the weight of the isolated object with low-amplitude vibration input. This is the environment of apparently stationary objects such as buildings (under normal conditions) or laboratory instruments.

Dynamic Loading

Dynamic loading involves accelerations and larger amplitude shock and vibration. This environment is present in vehicles, heavy machinery, and structures with significant movement.

- Cost:
    - Cost of providing isolation: Costs include the isolation system itself, whether it is a standard or custom product; a compressed air source if required; shipping from manufacturer to destination; installation; maintenance; and an initial vibration site survey to determine the need for isolation.
    - Relative costs of different isolation systems: Inexpensive shock mounts may need to be replaced due to dynamic loading cycles. A higher level of isolation which is effective at lower vibration frequencies and magnitudes generally costs more. Prices can range from a few dollars for bungee cords to millions of dollars for some space applications.
- Adjustment: Some isolation systems require manual adjustment to compensate for changes in weight load, weight distribution, temperature, and air pressure, whereas other systems are designed to automatically compensate for some or all of these factors.
- Maintenance: Some isolation systems are quite durable and require little or no maintenance. Others may require periodic replacement due to mechanical fatigue of parts or aging of materials.
- Size Constraints: The isolation system may have to fit in a restricted space in a laboratory or vacuum chamber, or within a machine housing.
- Nature of vibrations to be isolated or mitigated
    - Frequencies: If possible, it is important to know the frequencies of ambient vibrations. This can be determined with a site survey or accelerometer data processed through FFT analysis.
    - Amplitudes: The amplitudes of the vibration frequencies present can be compared with required levels to determine whether isolation is needed. In addition, isolators are designed for ranges of vibration amplitudes. Some isolators are not effective for very small amplitudes.
    - Direction: Knowing whether vibrations are horizontal or vertical can help to target isolation where it is needed and save money.
- Vibration specifications of item to be isolated: Many instruments or machines

have manufacturer-specified levels of vibration for the operating environment. The manufacturer may not guarantee the proper operation of the instrument if vibration exceeds the spec.

Comparison of Passive Isolators

Type of Passive Isolation	Applications	Typical Natural Frequency
Air Isolators	Large industrial equipment, some optics and instruments	1.5 - 3 Hz, large systems customized to 0.5 Hz
Springs or spring dampers	Heavy loads, pumps, compressors	3 - 9 Hz
Elastomer or cork pads	Large high-load applications where isolation of medium to high frequency noise and vibration is required	3 - 40 Hz, depending on size of pad and load
Molded or bonded elastomer mounts	Machinery, instruments, vehicles, aviation	10 - 20+ Hz
Negative-stiffness isolators	Electron microscopes, sensitive instruments, optics and laser systems, cryogenic systems	0.17 - 2.5 Hz
Wire rope isolators	Machinery, instruments, vehicles, aviation	10 - 40+ Hz
Bungee cord isolators	Laboratory, home, etc.	Depends on type of cord and the mass they support
Base isolators	Buildings and large structures	Low, seismic frequencies
Tuned Mass Dampers	Buildings, large structures, aerospace	Any, but usually used at low frequencies

Negative-Stiffness Vibration Isolator

Negative-Stiffness-Mechanism (NSM) vibration isolation systems offer a unique passive approach for achieving low vibration environments and isolation against sub-Hertz vibrations. “Snap-through” or “over-center” NSM devices are used to reduce the stiffness of elastic suspensions and create compact six-degree-of-freedom systems with low natural frequencies. Practical systems with vertical and horizontal natural frequencies as low as 0.2 to 0.5 Hz are possible. Electro-mechanical auto-adjust mechanisms compensate for varying weight loads and provide automatic leveling in multiple-isolator systems, similar to the function of leveling valves in pneumatic systems. All-metal systems can be configured which are compatible with high vacuums and other adverse environments such as high temperatures. These isolation systems enable vibration-sensitive instruments such as scanning probe microscopes, micro-hardness testers and scanning electron microscopes to operate in severe vibration environments sometimes encountered, for example, on upper floors of buildings and in clean rooms. Such operation would not be practical with pneumatic isolation systems. Similarly, they enable vibration-sensitive instruments to produce better images and data than those achievable with pneumatic isolators. The theory of operation of NSM vibration isolation systems is summarized, some typical systems and applications are described, and data on measured performance is presented. The theory of NSM isolation systems is explained in References 1 and 2. It is summarized briefly for convenience.

Vertical-Motion Isolation

A vertical-motion isolator is shown . It uses a conventional spring connected to an NSM consisting of two bars hinged at the center, supported at their outer ends on pivots, and loaded in compression by forces P. The spring is compressed by weight W to the operating position of the isolator, as shown in Figure. The stiffness of the isolator is K=KS-KN where KS is the spring stiffness and KN is the magnitude of a negative-stiffness which is a function of the length of the bars and the load P. The isolator stiffness can be made to approach zero while the spring supports the weight W.

Horizontal-Motion Isolation

A horizontal-motion isolator consisting of two beam-columns is illustrated in Figure. Each beam-column behaves like two fixed-free beam columns loaded axially by a weight load W. Without the weight load the beam-columns have horizontal stiffness KS With the weight load the lateral bending stiffness is reduced by the "beam-column" effect. This behavior is equivalent to a horizontal spring combined with an NSM so that the horizontal stiffness is K=KS-KN, and KN is the magnitude of the beam-column effect. Horizontal stiffness can be made to approach zero by loading the beam-columns to approach their critical buckling load.

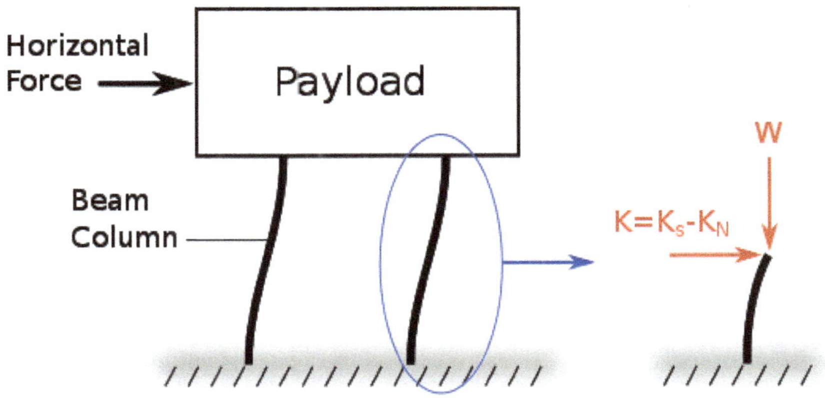

Beam column vibration isolation

Six-Degree-of-Freedom (six-DOF) Isolation

A six-DOF NSM isolator typically uses three isolators stacked in series: a tilt-motion isolator on top of a horizontal-motion isolator on top of a vertical-motion isolator. Figure shows a schematic of a vibration isolation system consisting of a weighted platform supported by a single six-DOF isolator incorporating the isolators. Flexures are used in place of the hinged bars. A tilt flexure serves as the tilt-motion isolator. A vertical-stiffness adjustment screw is used to adjust the compression force on the negative-stiffness flexures thereby changing the vertical stiffness. A vertical load adjustment screw is used to adjust for varying weight loads by raising or lowering the base of the support spring to keep the flexures in their straight, unbent operating positions.

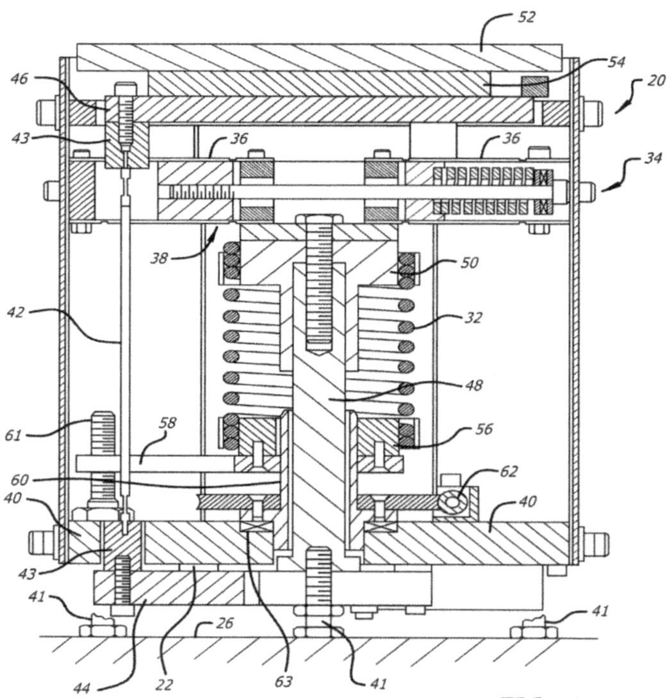

FIG. 1

Negative_Stiffness_Vibration_Isolator_Drawing

Vibration Isolation of Supporting Joint

The equipment or other mechanical components are necessarily linked to surrounding objects (the supporting joint - with the support; the unsupporting joint - the pipe duct or cable), thus presenting the opportunity for unwanted transmission of vibrations. Using a suitably designed vibration-isolator (absorber), vibration isolation of the supporting joint is realized. The accompanying illustration shows the attenuation of vibration levels, as measured before installation of the functioning gear on a vibration isolator as well as after installation, for a wide range of frequencies.

THE VIBRATION ISOLATOR

This is defined as a device that reflects and absorbs waves of oscillatory energy, extending from a piece of working machinery or electrical equipment, and with the desired effect being vibration insulation. The goal is to establish vibration isolation between a body transferring mechanical fluctuations and a supporting body (for example, between the machine and the foundation).

Boris Andreevich Shapiro, 1994 and 2006. Patent of Russia 2079419 and 2324086.

The illustration shows a vibration isolator from the series ВИ (~VI in Roman characters), as used in shipbuilding in Russia, for example the submarine St.Petersburg (Lada). The depicted ВИ devices allow loadings ranging from 5, 40 and 300 kg. They differ in their physical sizes, but all share the same fundamental design. The structure consists of a rubber envelope that is internally reinforced by a spring. During manufacture, the rubber and the spring are intimately and permanently connected as a result of the vulcanization process that is integral to the processing of the crude rubber material. Under action of weight loading of the machine, the rubber envelope deforms, and the spring is compressed or stretched. Therefore, in the direction of the spring›s cross section, twisting of the enveloping rubber occurs. The resulting elastic deformation of the rubber envelope results in very effective absorption of the vibration. This absorption is crucial to reliable vibration insulation, because it averts the potential for resonance effects. The amount of elastic deformation of the rubber largely dictates the magnitude of vibration absorption that can be attained; the entire device (including the spring itself) must be designed with this in mind. The design of the vibration isolator must also take into account potential exposure to shock loadings, in addition to the routine everyday vibrations. Lastly, the vibration isolator must also be designed for long-term durability as well as convenient integration into the environment in which it is to be used. Sleeves and flanges are typically employed in order to enable the vibration isolator to be securely fastened to the equipment and the supporting foundation.

Vibration Isolation of Unsupporting Joint

Vibration isolation of unsupporting joint is realized in the device named branch pipe a of isolating vibration.

Branch Pipe a of Isolating Vibration

Branch pipe a of isolating vibration is a part of a tube with elastic walls for reflection and absorption of waves of the oscillatory energy extending from the working pump over wall of the pipe duct. Is established between the pump and the pipe duct. On an illustration is presented the image a vibration-isolating branch pipe of a series ВИПБ. In a structure is used the rubber envelope, which is reinforced by a spring. Properties of an envelope are similar envelope to an isolator vibration. Has the device reducing axial effort from action of internal pressure up to zero.

Subframe Isolation

Another technique used to increase isolation is to use an isolated subframe. This splits the system with an additional mass/spring/damper system. This doubles the high frequency attenuation rolloff, at the cost of introducing additional low frequency modes which may cause the low frequency behavior to deteriorate.

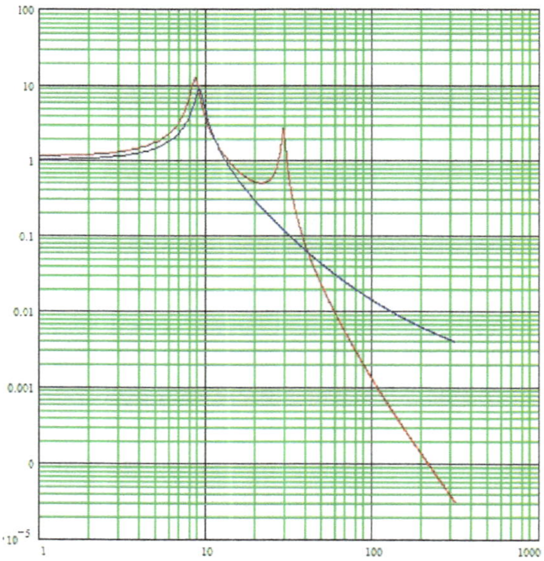

Subframe vibration isolation graph: force transmission on suspended body vs. frequency for rigidly and compliantly mounted subframes.

This is commonly used in the rear suspensions of cars with Independent Rear Suspension (IRS), and in the front subframes of some cars. The graph shows the force into the body for a subframe that is rigidly bolted to the body compared with the red curve that shows a compliantly mounted subframe. Above 42 Hz the compliantly mounted subframe is superior, but below that frequency the bolted in subframe is better.

ACTIVE ISOLATION

Active vibration isolation systems contain, along with the spring, a feedback circuit which consists of a sensor (for example a piezoelectric accelerometer or a geophone), a controller, and an actuator. The acceleration (vibration) signal is processed by a control circuit and amplifier. Then it feeds the electromagnetic actuator, which amplifies the signal. As a result of such a feedback system, a considerably stronger suppression of vibrations is achieved compared to ordinary damping. Active isolation today is used for applications where structures smaller than a micrometer have to be produced or measured. A couple of companies produce active isolation products as OEM for research, metrology, lithography and medical systems. Another important application is the semiconductor industry. In the microchip production, the smallest structures today are below 20 nm, so the machines which produce and check them have to oscillate much less.

Sensors for Active Isolation

- Piezoelectric accelerometers and force sensors
- MEM accelerometers
- Geophones
- Proximity sensors
- Interferometers

Geophone

A geophone is a device that converts ground movement (velocity) into voltage, which may be recorded at a recording station. The deviation of this measured voltage from the base line is called the seismic response and is analyzed for structure of the earth.

Geophone (SM-24), Bandwidth 10Hz to 240Hz, standard resistance 375Ω

Etymology

The term geophone derives from the Greek word "meaning "earth" and "phone" meaning "sound".

Construction

Geophones have historically been passive analog devices and typically comprise a spring-mounted magnetic mass moving within a wire coil to generate an electrical signal. Recent designs have been based on microelectromechanical systems (MEMS) technology which generates an electrical response to ground motion through an active feedback circuit to maintain the position of a small piece of silicon.

Geosource Inc. MD-79—8Hz, 335Ω geophone

The response of a coil/magnet geophone is proportional to ground velocity, while MEMS devices usually respond proportional to acceleration. MEMS have a much higher noise level (50 dB velocity higher) than geophones and can only be used in strong motion or active seismic applications.

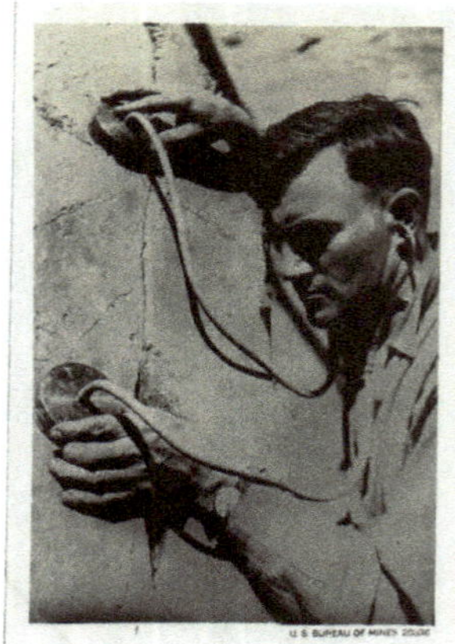

Man using a geophone

Frequency Response

The frequency response of a geophone is that of a harmonic oscillator, fully determined by corner frequency (typically around 10 Hz) and damping (typically 0.707). Since the corner frequency is proportional to the inverse root of the moving mass, geophones with low corner frequencies (< 1 Hz) become impractical. It is possible to lower the corner frequency electronically, at the price of higher noise and cost.

Although waves passing through the earth have a three-dimensional nature, geophones are normally constrained to respond to single dimension - usually the vertical. However, some applications require the full wave to be used and three-component or 3-C geophones are used. In analog devices, three moving coil elements are mounted in an orthogonal arrangement within a single case.

Uses

The majority of geophones are used in reflection seismology to record the energy waves reflected by the subsurface geology. In this case the primary interest is in the vertical motion of the Earth's surface. However, not all the waves are upwards travelling. A strong, horizontally transmitted wave known as ground-roll also generates vertical motion that can obliterate the weaker vertical signals. By using large areal arrays tuned to the wavelength of the ground-roll the dominant noise signals can be attenuated and the weaker data signals reinforced.

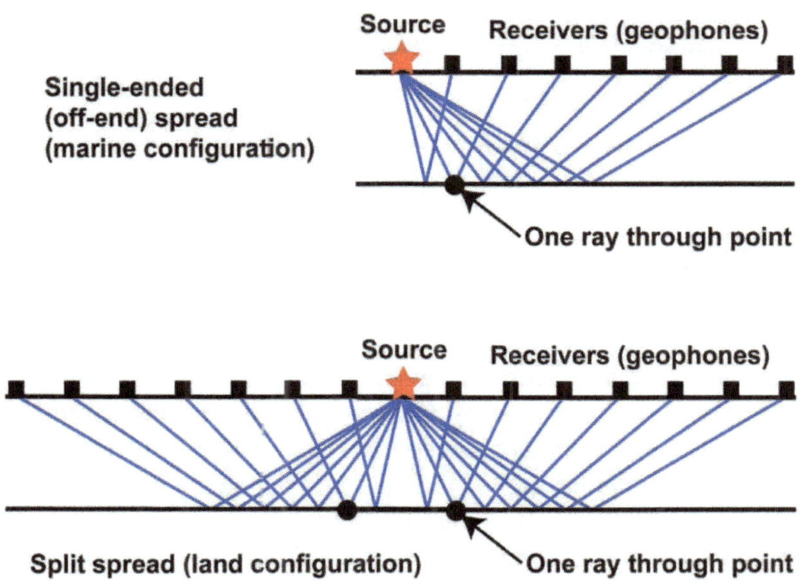

Single End Spread & Split Stread

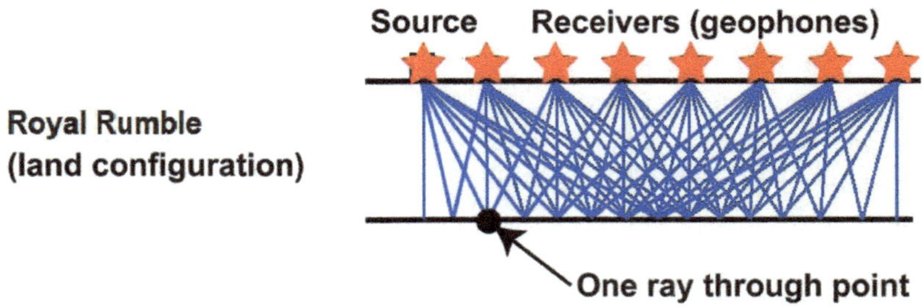

Royal Rumble

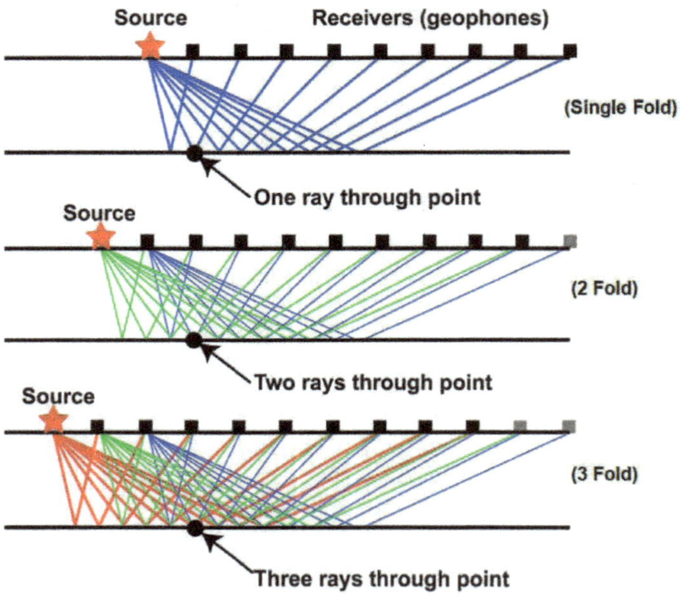

Source Receivers single fold-3 Fold

Analog geophones are very sensitive devices which can respond to very distant tremors. These small signals can be drowned by larger signals from local sources. It is possible though to recover the small signals caused by large but distant events by correlating signals from several geophones deployed in an array. Signals which are registered only at one or few geophones can be attributed to unwanted, local events and thus discarded. It can be assumed that small signals that register uniformly at all geophones in an array can be attributed to a distant and therefore significant event.

The sensitivity of passive geophones is typically 30 Volts/(meter/second), so they are in general not a replacement for broadband seismometers.

Conversely, some applications of geophones are interested only in very local events. A notable example is in the application of Remote Ground Sensors (RGS) incorporated in Unattended Ground Sensor (UGS) Systems. In such an application there is an area of interest which when penetrated a system operator is to be informed, perhaps by an alert which could be accompanied by supporting photographic data.

A network of geophones has been used on the moon surface too.

Proximity Sensor

A proximity sensor is a sensor able to detect the presence of nearby objects without any physical contact.

A proximity sensor often emits an electromagnetic field or a beam of electromagnetic radiation (infrared, for instance), and looks for changes in the field or return signal. The object being sensed is often referred to as the proximity sensor's target. Different proximity sensor targets demand different sensors. For example, a capacitive or photoelectric sensor might be suitable for a plastic target; an inductive proximity sensor always requires a metal target.

The maximum distance that this sensor can detect is defined "nominal range". Some sensors have adjustments of the nominal range or means to report a graduated detection distance. Some know these process as "thermosensation".

Infrared proximity sensor.

Proximity sensors can have a high reliability and long functional life because of the absence of mechanical parts and lack of physical contact between sensor and the sensed object.

Proximity sensors are commonly used on smartphones to detect (and skip) accidental touchscreen taps when held to the ear during a call. They are also used in machine vibration monitoring to measure the variation in distance between a shaft and its support bearing. This is common in large steam turbines, compressors, and motors that use sleeve-type bearings.

International Electrotechnical Commission (IEC) 60947-5-2 defines the technical details of proximity sensors.

A proximity sensor adjusted to a very short range is often used as a touch switch.

Air Touch and Gesture

Proximity sensors can be used to recognise air gestures and hover-manipulations. An array of proximity sensing elements can replace vision-camera or depth camera based solutions for the hand gesture detection. In particular, a car infotainment system (7 inch - 14 inch) in vehicle can employ the proximity sensors to cover the sensing area over the screen. For example, LG Electronics has recently filed a bunch of patents addressing this advanced technology.

Interferometry

Interferometry is a family of techniques in which waves, usually electromagnetic, are superimposed in order to extract information. Interferometry is an important investigative technique in the fields of astronomy, fiber optics, engineering metrology, optical metrology, oceanography, seismology, spectroscopy (and its applications to chemistry), quantum mechanics, nuclear and particle physics, plasma physics, remote sensing, biomolecular interactions, surface profiling, microfluidics, mechanical stress/strain measurement, velocimetry, and optometry.

Interferometers are widely used in science and industry for the measurement of small displacements, refractive index changes and surface irregularities. In analytical science, interferometers are used in continuous wave Fourier transform spectroscopy to analyze light containing features of absorption or emission associated with a substance or mixture. An astronomical interferometer consists of two or more separate telescopes that combine their signals, offering a resolution equivalent to that of a telescope of diameter equal to the largest separation between its individual elements.

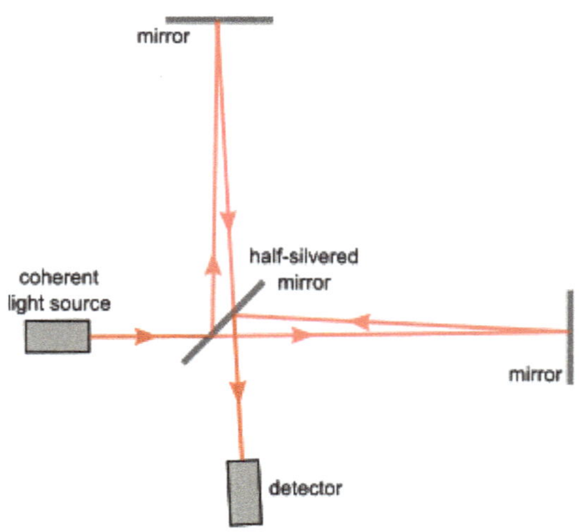

The light path through a Michelson interferometer. The two light rays with a common source combine at the half-silvered mirror to reach the detector. They may either interfere constructively (strengthening in intensity) if their light waves arrive in phase, or interfere destructively (weakening in intensity) if they arrive out of phase, depending on the exact distances between the three mirrors.

Basic Principles

Interferometry makes use of the principle of superposition to combine waves in a way that will cause the result of their combination to have some meaningful property that is diagnostic of the original state of the waves. This works because when two waves with the same frequency combine, the resulting intensity pattern is determined by the phase difference between the two waves—waves that are in phase will undergo constructive interference while waves that are out of phase will undergo destructive interference. Waves which are not completely in phase nor completely out of phase will have an intermediate intensity pattern, which can be used to determine their relative phase difference. Most interferometers use light or some other form of electromagnetic wave. [:3–12]

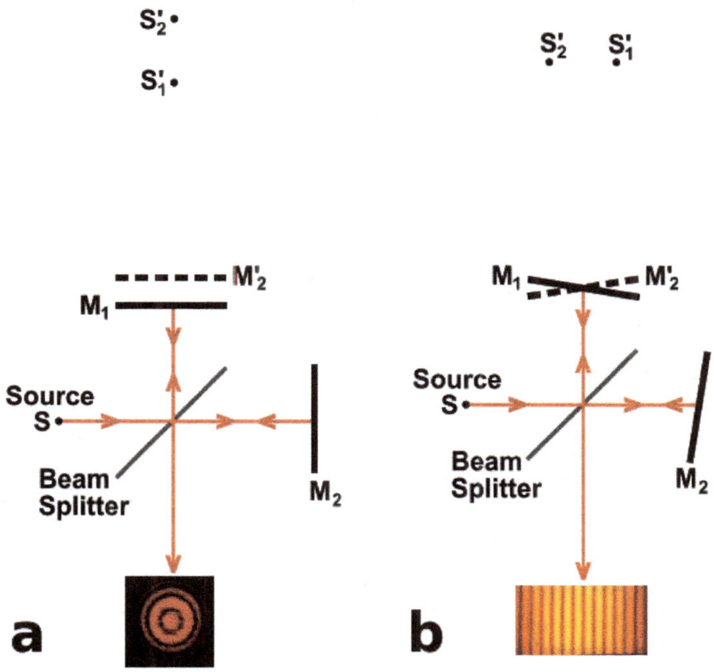

Formation of fringes in a Michelson interferometer

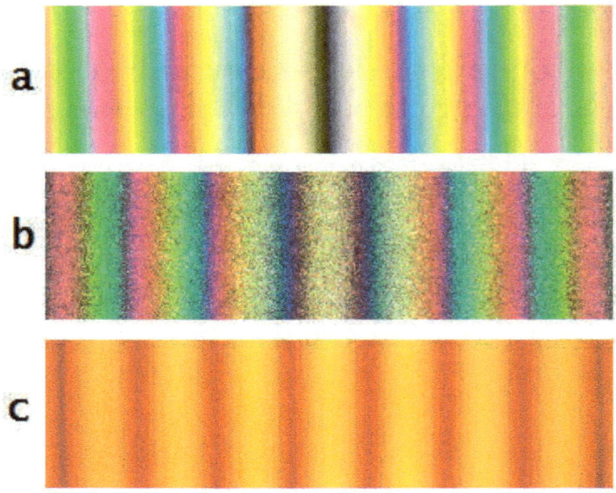

Colored and monochromatic fringes in a Michelson interferometer: (a) White light fringes where the two beams differ in the number of phase inversions; (b) White light fringes where the two beams have experienced the same number of phase inversions; (c) Fringe pattern using monochromatic light (sodium D lines)

Typically (the well-known Michelson configuration) a single incoming beam of coherent light will be split into two identical beams by a beam splitter (a partially reflecting mirror). Each of these beams travels a different route, called a path, and they are recombined before arriving at a detector. The path difference, the difference in the distance traveled by each beam, creates a phase difference between them. It is this introduced phase difference that creates the interference pattern between the initially identical waves. If a single beam has been split along two paths, then the phase difference is diagnostic of anything that changes the phase along the paths. This could be a physical change in the path length itself or a change in the refractive index along the path.

As seen in Figure, the observer has a direct view of mirror M_1 seen through the beam splitter, and sees a reflected image M'_2 of mirror M_2. The fringes can be interpreted as the result of interference between light coming from the two virtual images S'_1 and S'_2 of the original source S. The characteristics of the interference pattern depend on the nature of the light source and the precise orientation of the mirrors and beam splitter. In Figure, the optical elements are oriented so that S'_1 and S'_2 are in line with the observer, and the resulting interference pattern consists of circles centered on the normal to M_1 and M'_2. If, M_1 and M'_2 are tilted with respect to each other, the interference fringes will generally take the shape of conic sections (hyperbolas), but if M_1 and M'_2 overlap, the fringes near the axis will be straight, parallel, and equally spaced. If S is an extended source rather than a point source as illustrated, the fringes of

Use of white light will result in a pattern of colored fringes. The central fringe representing equal path length may be light or dark depending on the number of phase inversions experienced by the two beams as they traverse the optical system.

Categories

Interferometers and interferometric techniques may be categorized by a variety of criteria:

Homodyne versus Heterodyne Detection

In homodyne detection, the interference occurs between two beams at the same wavelength (or carrier frequency). The phase difference between the two beams results in a change in the intensity of the light on the detector. The resulting intensity of the light after mixing of these two beams is measured, or the pattern of interference fringes is viewed or recorded. Most of the interferometers discussed in this article fall into this category.

The heterodyne technique is used for (1) shifting an input signal into a new frequency range as well as (2) amplifying a weak input signal (assuming use of an active mixer). A weak input signal of frequency f_1 is mixed with a strong reference frequency f_2 from a local oscillator (LO). The nonlinear combination of the input signals creates two new signals, one at the sum $f_1 + f_2$ of the two frequencies, and the other at the difference $f_1 - f_2$. These new frequencies are called heterodynes. Typically only one of the new frequencies is desired, and the other signal is filtered out of the output of the mixer. The output signal will have an intensity proportional to the product of the amplitudes of the input signals.

The most important and widely used application of the heterodyne technique is in the superheterodyne receiver (superhet), invented by U.S. engineer Edwin Howard Armstrong in 1918. In this circuit, the incoming radio frequency signal from the antenna is mixed with a signal from a local oscillator (LO) and converted by the heterodyne technique to a lower fixed frequency signal called the intermediate frequency (IF). This IF is amplified and filtered, before being applied to a detector which extracts the audio signal, which is sent to the loudspeaker.

Optical heterodyne detection is an extension of the heterodyne technique to higher (visible) frequencies.

Double Path Versus Common Path

A double path interferometer is one in which the reference beam and sample beam travel along divergent paths. Examples include the Michelson interferometer, the Twyman-Green interferometer, and the Mach-Zehnder interferometer. After being perturbed by interaction with the sample under test, the sample beam is recombined with the reference beam to create an interference pattern which can then be interpreted.

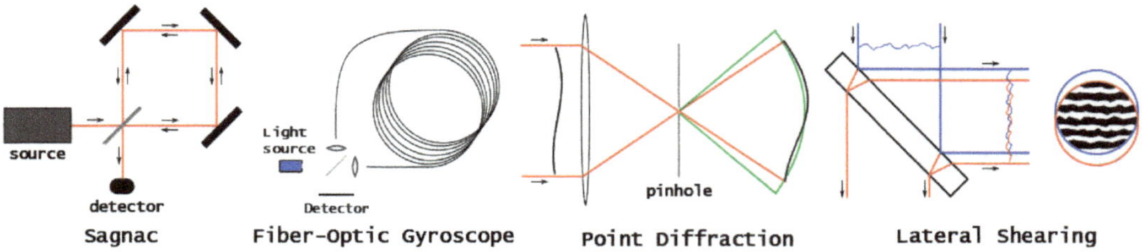

Four examples of common path interferometers

A common path interferometer is a class of interferometer in which the reference beam and sample beam travel along the same path. Figure illustrates the Sagnac interferometer, the fibre optic gyroscope, the point diffraction interferometer, and the lateral shearing interferometer. Other examples of common path interferometer include the Zernike phase contrast microscope, Fresnel's biprism, the zero-area Sagnac, and the scatterplate interferometer.

Wavefront Splitting Versus Amplitude Splitting

A wavefront splitting interferometer divides a light wavefront emerging from a point or a narrow slit (*i.e.* spatially coherent light) and, after allowing the two parts of the wavefront to travel through different paths, allows them to recombine. Figure illustrates Young's interference experiment and Lloyd's mirror. Other examples of wavefront splitting interferometer include the Fresnel biprism, the Billet Bi-Lens, and the Rayleigh interferometer.

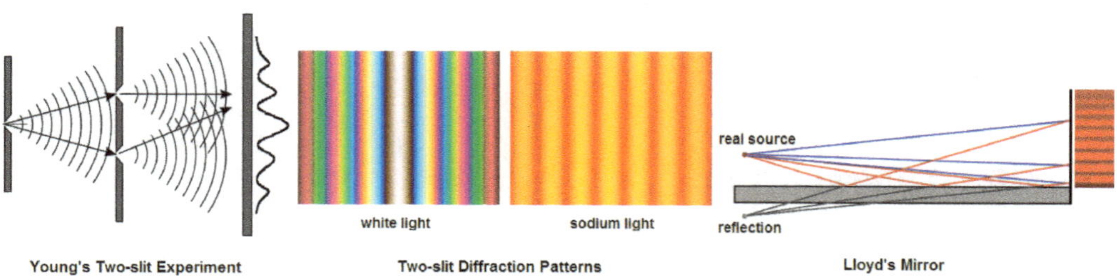

Two wavefront splitting interferometers

In 1803, Young's interference experiment played a major role in the general acceptance of the wave theory of light. If white light is used in Young's experiment, the result is a white central band of constructive interference corresponding to equal path length from the two slits, surrounded by a symmetrical pattern of colored fringes of diminishing intensity. In addition to continuous electromagnetic radiation, Young's experiment has been performed with individual photons, with electrons, and with buckyballmolecules large enough to be seen under an electron microscope.

Lloyd's mirror generates interference fringes by combining direct light from a source (blue lines) and light from the source's reflected image (red lines) from a mirror held at grazing incidence. The result is an asymmetrical pattern of fringes. Interestingly, the band of equal path length, nearest the mirror, is dark rather than bright. In 1834, Humphrey Lloyd interpreted this effect as proof that the phase of a front-surface reflected beam is inverted.

An amplitude splitting interferometer uses a partial reflector to divide the amplitude of the incident wave into separate beams which are separated and recombined. Figure illustrates the Fizeau, Mach–Zehnder and Fabry–Pérot interferometers. Other examples of amplitude splitting interferometer include the Michelson, Twyman–Green, Laser Unequal Path, and Linnik interferometer.

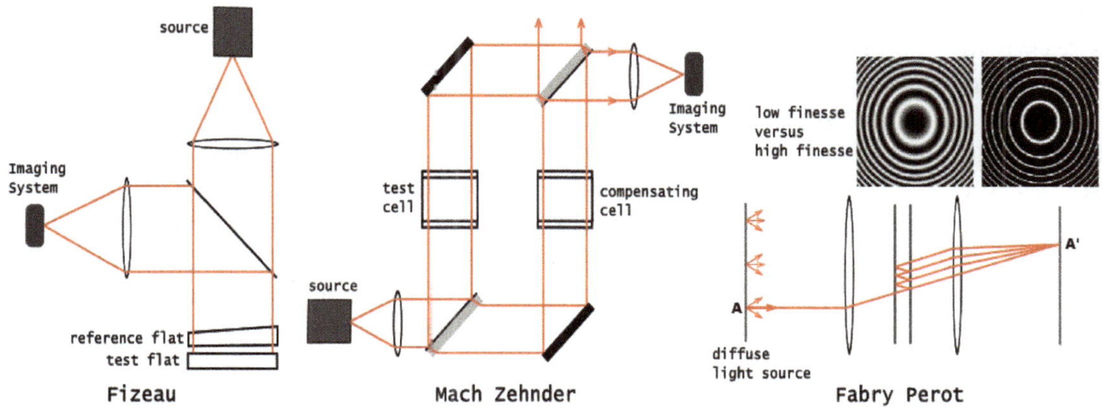

Three amplitude-splitting interferometers: Fizeau, Mach–Zehnder, and Fabry Pérot

The Fizeau interferometer is shown as it might be set up to test an optical flat. A precisely figured reference flat is placed on top of the flat being tested, separated by narrow spacers. The reference flat is slightly beveled (only a fraction of a degree of beveling is necessary) to prevent the rear surface of the flat from producing interference fringes. Separating the test and reference flats allows the two flats to be tilted with respect to each other. By adjusting the tilt, which adds a controlled phase gradient to the fringe pattern, one can control the spacing and direction of the fringes, so that one may obtain an easily interpreted series of nearly parallel fringes rather than a complex swirl of contour lines. Separating the plates, however, necessitates that the illuminating light be collimated. Fig 6 shows a collimated beam of monochromatic light illuminating the two flats and a beam splitter allowing the fringes to be viewed on-axis.

The Mach–Zehnder interferometer is a more versatile instrument than the Michelson interferometer. Each of the well separated light paths is traversed only once, and the fringes can be adjusted so that they are localized in any desired plane. Typically, the fringes would be adjusted to lie in the same plane as the test object, so that fringes and test object can be photographed together. If it is decided to produce fringes in white light, then, since white light has a limited coherence length, on the order of micrometers, great care must be taken to equalize the optical paths or no fringes will be visible. As illustrated in Figure, a compensating cell would be placed in the path of the reference beam to match the test cell. Note also the precise orientation of the beam splitters. The reflecting surfaces of the beam splitters would be oriented so that the test and reference beams pass through an equal amount of glass. In this orientation, the test and reference beams each experience two front-surface reflections, resulting in the same number of phase inversions. The result is that light traveling an equal optical path length in the test and reference beams produces a white light fringe of constructive interference.

The heart of the Fabry–Pérot interferometer is a pair of partially silvered glass optical flats spaced several millimeters to centimeters apart with the silvered surfaces facing each other. (Alternatively, a Fabry–Pérot *etalon* uses a transparent plate with two parallel reflecting surfaces.) As with the Fizeau interferometer, the flats are slightly beveled. In a typical system, illumination is provided by a diffuse source set at the focal plane of a collimating lens. A focusing lens produces what would be an inverted image of the source if the paired flats were not present; *i.e.* in the absence of the paired flats, all light emitted from point A passing through the optical system would be focused at point A'. In Figure, only one ray emitted from point A on the source is traced. As the ray passes through the paired flats, it is multiply reflected to produce multiple transmitted rays which are collected by the focusing lens and brought to point A' on the screen. The complete interference pattern takes the appearance of a set of concentric rings. The sharpness of the rings depends on the reflectivity of the flats. If the reflectivity is high, resulting in a high Q factor (*i.e.* high finesse), monochromatic light produces a set of narrow bright rings against a dark background. The low-finesse image corresponds to a reflectivity of 0.04 (*i.e.* unsilvered surfaces) *versus* a reflectivity of 0.95 for the high-finesse image.

Michelson and Morley (1887) and other early experimentalists using interferometric techniques in an attempt to measure the properties of the luminiferous aether, used monochromatic light only for initially setting up their equipment, always switching to white light for the actual measurements. The reason is that measurements were recorded visually. Monochromatic light would result in a uniform fringe pattern. Lacking modern means of environmental temperature control, experimentalists struggled with continual fringe drift even though the interferometer might be set up in a basement. Since the fringes would occasionally disappear due to vibrations by passing horse traffic, distant thunderstorms and the like, it would be easy for an observer to "get lost" when the fringes returned to visibility. The advantages of white light, which produced a distinctive colored fringe pattern, far outweighed the difficulties of aligning the apparatus due to its low coherence length. This was an early example of the use of white light to resolve the "2 pi ambiguity".

Applications

Physics and astronomy

In physics, one of the most important experiments of the late 19th century was the famous "failed experiment" of Michelson and Morley which provided evidence for special relativity. Recent repetitions of the Michelson–Morley experiment perform heterodyne measurements of beat frequencies of crossed cryogenic optical resonators. Figure illustrates a resonator experiment performed by Müller et al. in 2003. Two optical resonators constructed from crystalline sapphire, controlling the frequencies of two lasers, were set at right angles within a helium cryostat. A frequency comparator measured the beat frequency of the combined outputs of the two resonators. As of 2009, the precision by which anisotropy of the speed of light can be excluded in resonator experiments is at the 10^{-17} level.

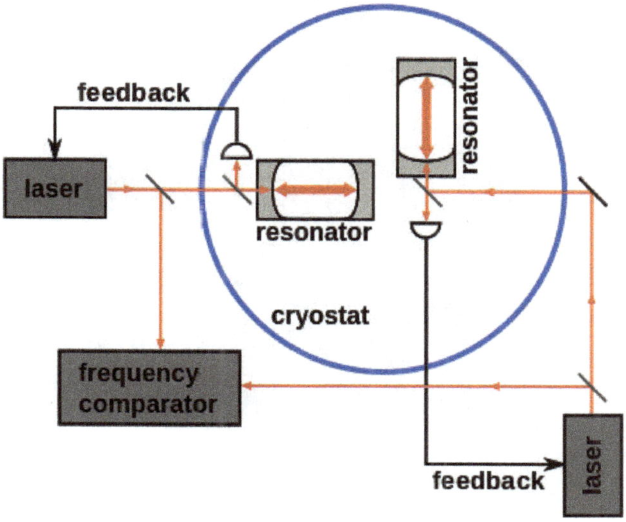

User:Stigmatella aurantiaca

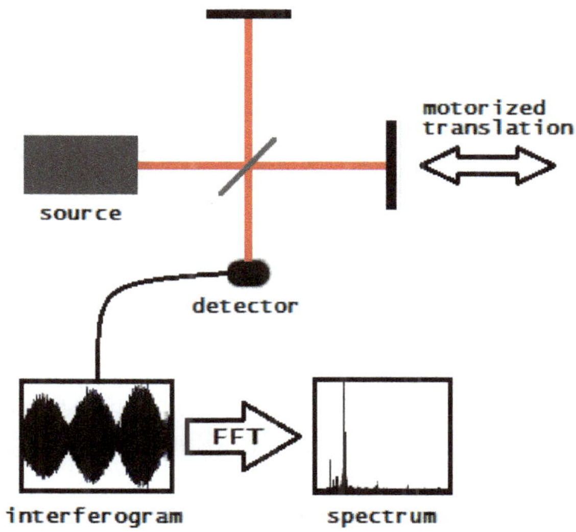

A Fourier transform spectrometer comprises a Michelson interferometer with a movable mirror. A Fourier transform converts the interferogram into the actual spectrum.

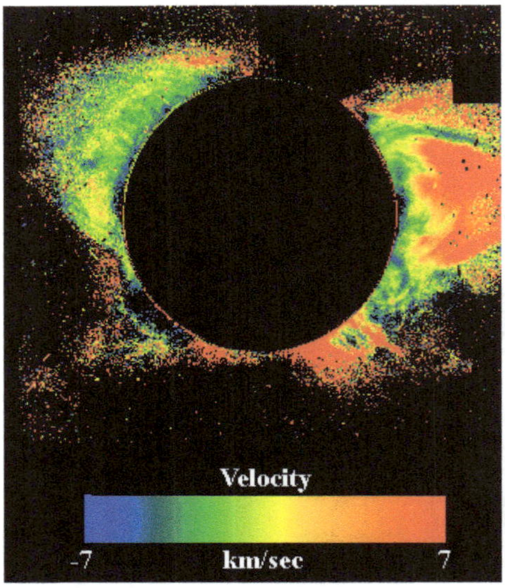

LASCO C1a

Michelson interferometers are used in tunable narrow band optical filters and as the core hardware component of Fourier transform spectrometers.

When used as a tunable narrow band filter, Michelson interferometers exhibit a number of advantages and disadvantages when compared with competing technologies such as Fabry–Pérot interferometers or Lyot filters. Michelson interferometers have the largest field of view for a specified wavelength, and are relatively simple in operation, since tuning is via mechanical rotation of waveplates rather than via high voltage control of piezoelectric crystals or lithium niobate optical modulators as used in a Fabry–Pérot system. Compared with Lyot filters, which use birefringent elements, Michelson interferometers have a relatively low temperature sensitivity. On the negative side, Michelson interferometers have a relatively restricted wavelength range and require use of prefilters which restrict transmittance.

Illustrates the operation of a Fourier transform spectrometer, which is essentially a Michelson interferometer with one mirror movable. (A practical Fourier transform spectrometer would substitute corner cube reflectors for the flat mirrors of the conventional Michelson interferometer, but for simplicity, the illustration does not show this.) An interferogram is generated by making measurements of the signal at many discrete positions of the moving mirror. A Fourier transform converts the interferogram into an actual spectrum.

Figure shows a doppler image of the solar corona made using a tunable Fabry-Pérot interferometer to recover scans of the solar corona at a number of wavelengths near the FeXIV green line. The picture is a color-coded image of the doppler shift of the line, which may be associated with the coronal plasma velocity towards or away from the satellite camera.

Fabry-Pérot thin-film etalons are used in narrow bandpass filters capable of selecting a single spectral line for imaging; for example, the H-alpha line or the Ca-K line of the Sun or stars. Figure shows an Extreme ultraviolet Imaging Telescope (EIT) image of the

Sun at 195 **Ångströms**, corresponding to a spectral line of multiply-ionized iron atoms. EIT used multilayer coated reflective mirrors that were coated with alternate layers of a light "spacer" element (such as silicon), and a heavy "scatterer" element (such as molybdenum). Approximately 100 layers of each type were placed on each mirror, with a thickness of around 10 nm each. The layer thicknesses were tightly controlled so that at the desired wavelength, reflected photons from each layer interfered constructively.

The Laser Interferometer Gravitational-Wave Observatory (LIGO) uses two 4-km Michelson-Fabry-Pérot interferometers for the detection of gravitational waves. In this application, the Fabry–Pérot cavity is used to store photons for almost a millisecond while they bounce up and down between the mirrors. This increases the time a gravitational wave can interact with the light, which results in a better sensitivity at low frequencies. Smaller cavities, usually called mode cleaners, are used for spatial filtering and frequency stabilization of the main laser. The first observation of gravitational waves occurred on September 14, 2015.

The Mach-Zehnder interferometer's relatively large and freely accessible working space, and its flexibility in locating the fringes has made it the interferometer of choice for visualizing flow in wind tunnels, and for flow visualization studies in general. It is frequently used in the fields of aerodynamics, plasma physics and heat transfer to measure pressure, density, and temperature changes in gases.

Mach-Zehnder interferometers are also used to study one of the most counterintuitive predictions of quantum mechanics, the phenomenon known as quantum entanglement.

The VLA interferometer

An astronomical interferometer achieves high-resolution observations using the technique of aperture synthesis, mixing signals from a cluster of comparatively small telescopes rather than a single very expensive monolithic telescope.

Early radio telescope interferometers used a single baseline for measurement. Later astronomical interferometers, such as the Very Large Array illustrated in Fig 11, used arrays of telescopes arranged in a pattern on the ground. A limited number of baselines will result in insufficient coverage. This was alleviated by using the rotation of the Earth to rotate the array relative to the sky. Thus, a single baseline could measure information in multiple orientations by taking repeated measurements, a technique called *Earth-rotation synthesis*. Baselines thousands of kilometers long were achieved using very long baseline interferometry.

ALMA is an astronomical interferometer located in Chajnantor Plateau

Astronomical optical interferometry has had to overcome a number of technical issues not shared by radio telescope interferometry. The short wavelengths of light necessitate extreme precision and stability of construction. For example, spatial resolution of 1 milliarcsecond requires 0.5 **μm** stability in a 100 m baseline. Optical interferometric measurements require high sensitivity, low noise detectors that did not become available until the late 1990s. Astronomical seeing, the turbulence that causes stars to twinkle, introduces rapid, random phase changes in the incoming light, requiring kilohertz data collection rates to be faster than the rate of turbulence. Despite these technical difficulties, roughly a dozen astronomical optical interferometers are now in operation offering resolutions down to the fractional milliarcsecond range. This linked video shows a movie assembled from aperture synthesis images of the Beta Lyrae system, a binary star system approximately 960 light-years (290 parsecs) away in the constellation Lyra, as observed by the CHARA array with the MIRC instrument. The brighter component is the primary star, or the mass donor. The fainter component is the thick disk surrounding the secondary star, or the mass gainer. The two components are separated by 1 milli-arcsecond. Tidal distortions of the mass donor and the mass gainer are both clearly visible.

The wave character of matter can be exploited to build interferometers. The first examples of matter interferometers were electron interferometers, later followed by neutron interferometers. Around 1990 the first atom interferometers were demonstrated, later followed by interferometers employing molecules.

Electron holography is an imaging technique that photographically records the electron interference pattern of an object, which is then reconstructed to yield a greatly magnified image of the original object. This technique was developed to enable greater resolution in electron microscopy than is possible using conventional imaging techniques. The resolution of conventional electron microscopy is not limited by electron wavelength, but by the large aberrations of electron lenses.

Neutron interferometry has been used to investigate the Aharonov–Bohm effect, to examine the effects of gravity acting on an elementary particle, and to demonstrate a strange behavior of fermions that is at the basis of the Pauli exclusion principle: Unlike macroscopic objects, when fermions are rotated by 360° about any axis, they do not return to their original state, but develop a minus sign in their wave function. In other words, a fermion needs to be rotated 720° before returning to its original state.

Atom interferometry techniques are reaching sufficient precision to allow laboratory-scale tests of general relativity.

Interferometers are used in atmospheric physics for high-precision measurements of trace gases via remote sounding of the atmosphere. There are several examples of interferometers that utilize either absorption or emission features of trace gases. A typical use would be in continual monitoring of the column concentration of trace gases such as ozone and carbon monoxide above the instrument.

Engineering and Applied Science

Newton (test plate) interferometry is frequently used in the optical industry for testing the quality of surfaces as they are being shaped and figured. Figure shows photos of reference flats being used to check two test flats at different stages of completion, showing the different patterns of interference fringes. The reference flats are resting with their bottom surfaces in contact with the test flats, and they are illuminated by a monochromatic light source. The light waves reflected from both surfaces interfere, resulting in a pattern of bright and dark bands.

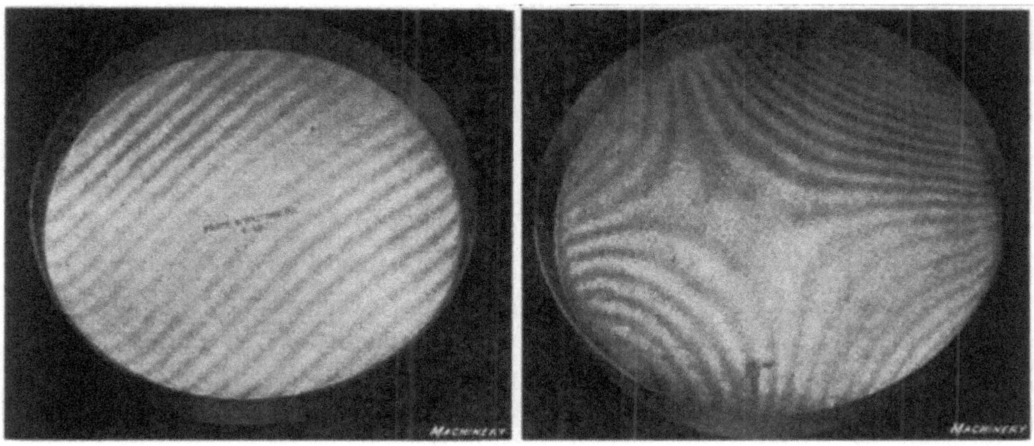

Optical flat interference fringes

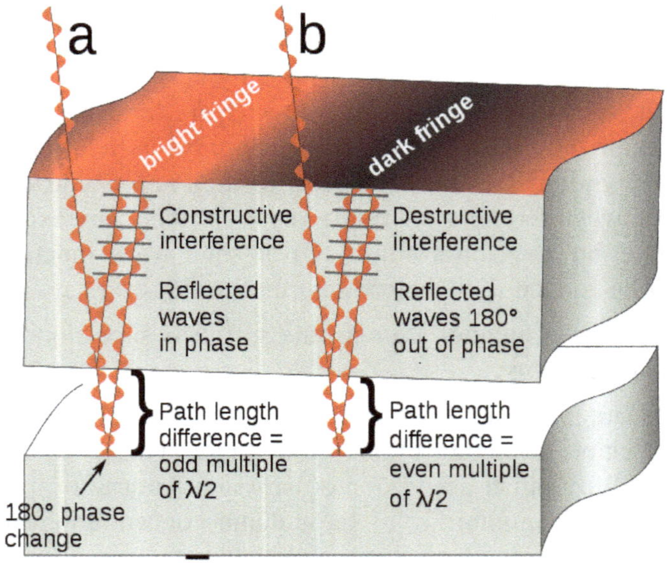

How interference fringes are formed by an optical flat resting on a reflective surface. The gap between the surfaces and the wavelength of the light waves are greatly exaggerated.

The surface in the left photo is nearly flat, indicated by a pattern of straight parallel interference fringes at equal intervals. The surface in the right photo is uneven, resulting in a pattern of curved fringes. Each pair of adjacent fringes represents a difference in surface elevation of half a wavelength of the light used, so differences in elevation can be measured by counting the fringes. The flatness of the surfaces can be measured to millionths of an inch by this method. To determine whether the surface being tested is concave or convex with respect to the reference optical flat, any of several procedures may be adopted. One can observe how the fringes are displaced when one presses gently on the top flat. If one observes the fringes in white light, the sequence of colors becomes familiar with experience and aids in interpretation. Finally one may compare the appearance of the fringes as one moves ones head from a normal to an oblique viewing position. These sorts of maneuvers, while common in the optical shop, are not suitable in a formal testing environment. When the flats are ready for sale, they will typically be mounted in a Fizeau interferometer for formal testing and certification.

Fabry-Pérot etalons are widely used in telecommunications, lasers and spectroscopy to control and measure the wavelengths of light. Dichroic filters are multiple layer thin-film etalons. In telecommunications, wavelength-division multiplexing, the technology that enables the use of multiple wavelengths of light through a single optical fiber, depends on filtering devices that are thin-film etalons. Single-mode lasers employ etalons to suppress all optical cavity modes except the single one of interest. [:42]

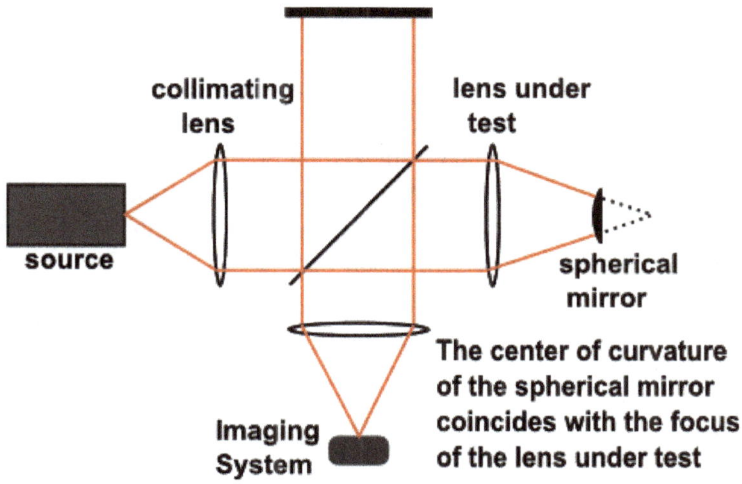

Twyman-Green Interferometer

The Twyman–Green interferometer, invented by Twyman and Green in 1916, is a variant of the Michelson interferometer widely used to test optical components. The basic characteristics distinguishing it from the Michelson configuration are the use of a monochromatic point light source and a collimator. It is interesting to note that Michelson (1918) criticized the Twyman-Green configuration as being unsuitable for the testing of large optical components, since the light sources available at the time had limited coherence length. Michelson pointed out that constraints on geometry forced by limited coherence length required the use of a reference mirror of equal size to the test mirror, making the Twyman-Green impractical for many purposes. Decades later, the advent of laser light sources answered Michelson's objections. (A Twyman-Green interferometer using a laser light source and unequal path length is known as a Laser Unequal Path Interferometer, or LUPI.) Figure illustrates a Twyman-Green interferometer set up to test a lens. Light from a monochromatic point source is expanded by a diverging lens (not shown), then is collimated into a parallel beam. A convex spherical mirror is positioned so that its center of curvature coincides with the focus of the lens being tested. The emergent beam is recorded by an imaging system for analysis.

Mach-Zehnder interferometers are being used in integrated optical circuits, in which light interferes between two branches of a waveguide that are externally modulated to vary their relative phase. A slight tilt of one of the beam splitters will result in a path difference and a change in the interference pattern. Mach-Zehnder interferometers are the basis of a wide variety of devices, from RF modulators to sensors to optical switches.

The latest proposed extremely large astronomical telescopes, such as the Thirty Meter Telescope and the European Extremely Large Telescope, will be of segmented design. Their primary mirrors will be built from hundreds of hexagonal mirror segments. Polishing and figuring these highly aspheric and non-rotationally symmetric mirror segments presents a major challenge. Traditional means of optical testing compares a surface against a spherical reference with the aid of a null corrector. In recent years, computer-generated holograms (CGHs) have begun to supplement null correctors in test setups for complex aspheric surfaces. Figure illustrates how this is done. Unlike the figure, actual CGHs have line spacing on the order of 1 to 10 **μm.** When laser light is passed through the CGH, the zero-order diffracted beam experiences no wavefront modification. The wavefront of the first-order diffracted beam, however, is modified to match the desired shape of the test surface. In the illustrated Fizeau interferometer test setup, the zero-order diffracted beam is directed towards the spherical reference surface, and the first-order diffracted beam is directed towards the test surface in such a way that the two reflected beams combine to form interference fringes. The same test setup can be used for the innermost mirrors as for the outermost, with only the CGH needing to be exchanged.

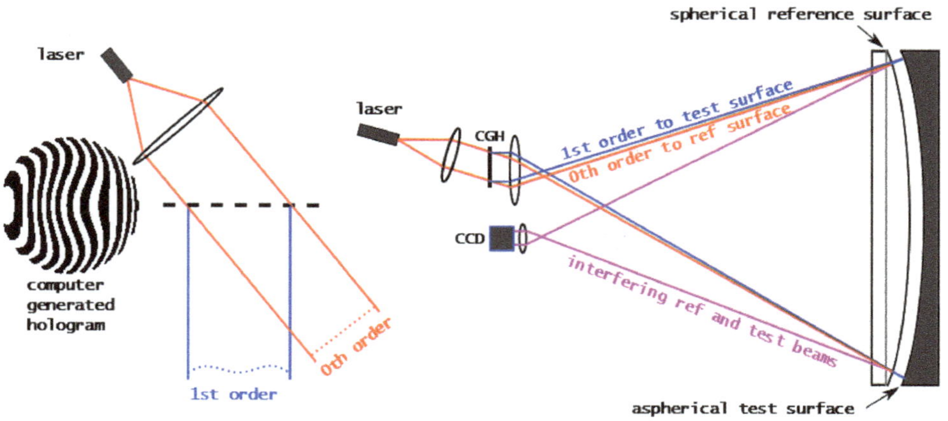

Optical testing with a Fizeau interferometer and a computer generated hologram

Ring laser gyroscopes (RLGs) and fibre optic gyroscopes (FOGs) are interferometers used in navigation systems. They operate on the principle of the Sagnac effect. The distinction between RLGs and FOGs is that in a RLG, the entire ring is part of the laser while in a FOG, an external laser injects counter-propagating beams into an optical fiber ring, and rotation of the system then causes a relative phase shift between those beams. In a RLG, the observed phase shift is proportional to the accumulated rotation, while in a FOG, the observed phase shift is proportional to the angular velocity.

In telecommunication networks, heterodyning is used to move frequencies of individual signals to different channels which may share a single physical transmission line. This is called frequency division multiplexing (FDM). For example, a coaxial cable used by a cable television system can carry 500 television channels at the same time because each one is given a different frequency, so they don't interfere with one another. Continuous wave (CW) doppler radar detectors are basically heterodyne detection devices that compare transmitted and reflected beams.

Optical heterodyne detection is used for coherent Doppler lidar measurements capable of detecting very weak light scattered in the atmosphere and monitoring wind speeds with high accuracy. It has application in optical fiber communications, in various high resolution spectroscopic techniques, and the self-heterodyne method can be used to measure the linewidth of a laser.

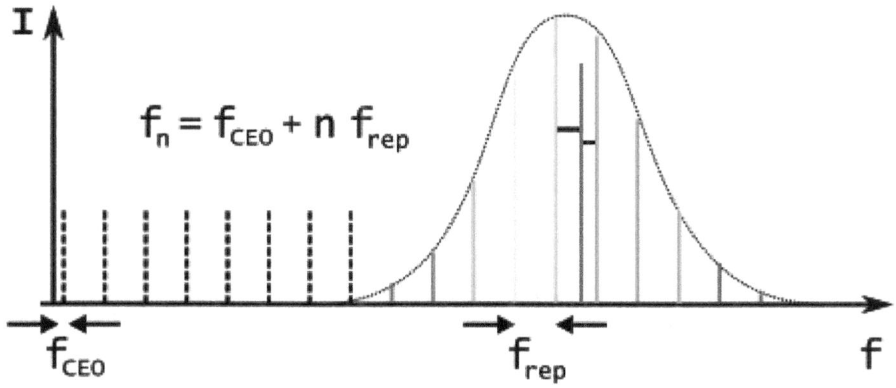

Frequency comb of a mode-locked laser. The dashed lines represent an extrapolation of the mode frequencies towards the frequency of the carrier–envelope offset (CEO). The vertical grey line represents an unknown optical frequency. The horizontal black lines indicate the two lowest beat frequency measurements.

Optical heterodyne detection is an essential technique used in high-accuracy measurements of the frequencies of optical sources, as well as in the stabilization of their frequencies. Until a relatively few years ago, lengthy frequency chains were needed to connect the microwave frequency of a cesium or other atomic time source to optical frequencies. At each step of the chain, a frequency multiplier would be used to produce a harmonic of the frequency of that step, which would be compared by heterodyne detection with the next step (the output of a microwave source, far infrared laser, infrared laser, or visible laser). Each measurement of a single spectral line required several years of effort in the construction of a custom frequency chain. Currently, optical frequency combs have provided a much simpler method of measuring optical frequencies. If a mode-locked laser is modulated to form a train of pulses, its spectrum is seen to consist of the carrier frequency surrounded by a closely spaced comb of optical sideband frequencies with a spacing equal to the pulse repetition frequency. The pulse repetition frequency is locked to that of the frequency standard, and the frequencies of the comb elements at the red end of the spectrum are doubled and heterodyned with the frequencies of the comb elements at the blue end of the spectrum, thus allowing the comb to serve as its own reference. In this manner, locking of the frequency comb output to an atomic standard can be performed in a single step. To measure an unknown frequency, the frequency comb output is dispersed into a spectrum. The unknown frequency is overlapped with the appropriate spectral segment of the comb and the frequency of the resultant heterodyne beats is measured.

One of the most common industrial applications of optical interferometry is as a versatile measurement tool for the high precision examination of surface topography. Popular interferometric measurement techniques include Phase Shifting Interferometry (PSI), and , also known as scanning white light interferometry (SWLI) or by the ISO term Coherence Scanning Interferometry (CSI), CSI exploits coherence to extend the range of capabilities for interference microscopy. These techniques are widely used in micro-electronic and micro-optic fabrication. PSI uses monochromatic light and provides very precise measurements; however it is only usable for surfaces that are very smooth. CSI often uses white light and high numerical apertures, and rather than looking at the phase of the fringes, as does PSI, looks for best position of maximum fringe contrast or some other feature of the overall fringe pattern. In its simplest form, CSI provides less precise measurements than PSI but can be used on rough surfaces. Some configurations of CSI, variously known as and Enhanced VSI (EVSI), high-resolution SWLI or Frequency Domain Analysis (FDA), use coherence effects in combination with interference phase to enhance precision.

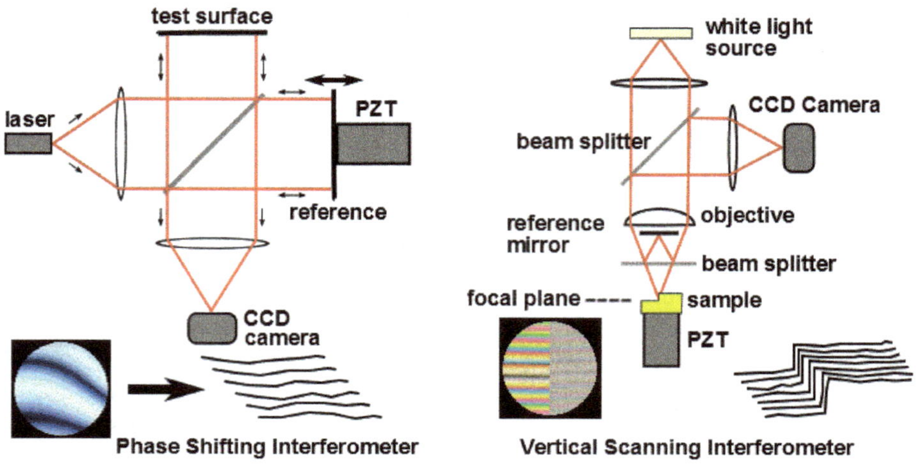

Phase shifting and Coherence scanning interferometers

Phase Shifting Interferometry addresses several issues associated with the classical analysis of static interferograms. Classically, one measures the positions of the fringe centers. As seen in Figure, fringe deviations from straightness and equal spacing provide a measure of the aberration. Errors in determining the location of the fringe centers provide the inherent limit to precision of the classical analysis, and any intensity variations across the interferogram will also introduce error. There is a trade-off between precision and number of data points: closely spaced fringes provide many data points of low precision, while widely spaced fringes provide a low number of high precision data points. Since fringe center data is all that one uses in the classical analysis, all of the other information that might theoretically be obtained by detailed analysis of the intensity variations in an interferogram is thrown away. Finally, with static interferograms, additional information is needed to determine the polarity of the wavefront: In Figure, one can see that the tested surface on the right deviates from flatness, but one cannot tell from this single image whether this deviation from flatness is concave or convex. Traditionally, this information would be obtained using non-automated means, such as by observing the direction that the fringes move when the reference surface is pushed.

Phase shifting interferometry overcomes these limitations by not relying on finding fringe centers, but rather by collecting intensity data from every point of the CCD image sensor. As seen in Figure, multiple interferograms (at least three) are analyzed with the reference optical surface shifted by a precise fraction of a wavelength between each exposure using a piezoelectric transducer (PZT). Alternatively, precise phase shifts can be introduced by modulating the laser frequency. The captured images are processed by a computer to calculate the optical wavefront errors. The precision and reproducibility of PSI is far greater than possible in static interferogram analysis, with measurement repeatabilities of a hundredth of a wavelength being routine. Phase shifting technology has been adapted to a variety of interferometer types such as Twyman-Green, Mach–Zehnder, laser Fizeau, and even common path configurations such as point diffraction and lateral shearing interferometers. More generally, phase shifting techniques can be adapted to almost any system that uses fringes for measurement, such as holographic and speckle interferometry.

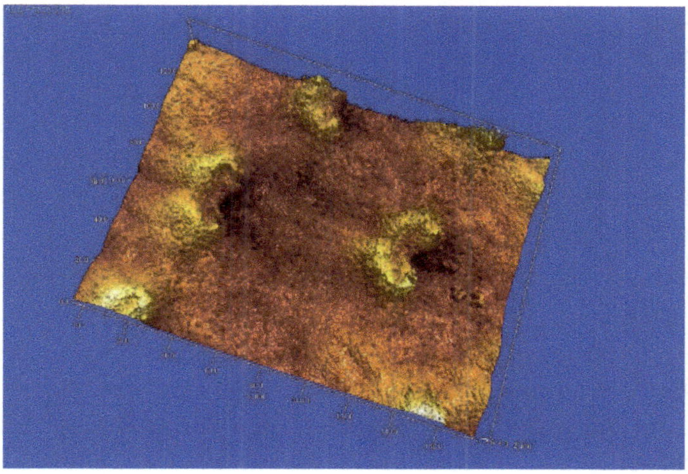

Lunate cells of *Nepenthes khasiana* visualized by Scanning White Light Interferometry (SWLI)

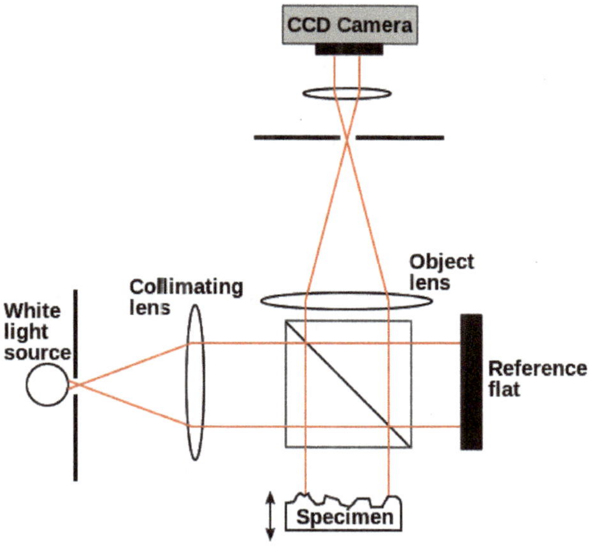

Twyman-Green interferometer set up as a white light scanner

In coherence scanning interferometry, interference is only achieved when the path length delays of the interferometer are matched within the coherence time of the light source. CSI monitors the fringe contrast rather than the phase of the fringes. Figure illustrates a CSI microscope using a Mirau interferometer in the objective; other forms of interferometer used with white light include the Michelson interferometer (for low magnification objectives, where the reference mirror in a Mirau objective would interrupt too much of the aperture) and the Linnik interferometer (for high magnification objectives with limited working distance). The sample (or alternatively, the objective) is moved vertically over the full height range of the sample, and the position of maximum fringe contrast is found for each pixel. The chief benefit of coherence scanning interferometry is that systems can be designed that do not suffer from the 2 pi ambiguity of coherent interferometry, and as seen in Figure, which scans a 180μm x 140μm x 10μm volume, it is well suited to profiling steps and rough surfaces. The axial resolution of the system is determined in part by the coherence length of the light source. Industrial applications include in-process surface metrology, roughness measurement, 3D surface metrology in hard-to-reach spaces and in hostile environments, profilometry of surfaces with high aspect ratio features (grooves, channels, holes), and film thickness measurement (semi-conductor and optical industries, etc.).

Figure illustrates a Twyman–Green interferometer set up for white light scanning of a macroscopic object.

Holographic interferometry is a technique which uses holography to monitor small deformations in single wavelength implementations. In multi-wavelength implementations, it is used to perform dimensional metrology of large parts and assemblies and to detect larger surface defects.

Holographic interferometry was discovered by accident as a result of mistakes committed during the making of holograms. Early lasers were relatively weak and photographic plates were insensitive, necessitating long exposures during which vibrations or minute shifts might occur in the optical system. The resultant holograms, which showed the holographic subject covered with fringes, were considered ruined.

Eventually, several independent groups of experimenters in the mid-60s realized that the fringes encoded important information about dimensional changes occurring in the subject, and began intentionally producing holographic double exposures. The main Holographic interferometry article covers the disputes over priority of discovery that occurred during the issuance of the patent for this method.

Double- and multi- exposure holography is one of three methods used to create holographic interferograms. A first exposure records the object in an unstressed state. Subsequent exposures on the same photographic plate are made while the object is subjected to some stress. The composite image depicts the difference between the stressed and unstressed states.

Real-time holography is a second method of creating holographic interferograms. A holograph of the unstressed object is created. This holograph is illuminated with a reference beam to generate a hologram image of the object directly superimposed over the original object itself while the object is being subjected to some stress. The object waves from this hologram image will interfere with new waves coming from the object.

This technique allows real time monitoring of shape changes. The third method, time-average holography, involves creating a holograph while the object is subjected to a periodic stress or vibration. This yields a visual image of the vibration pattern.

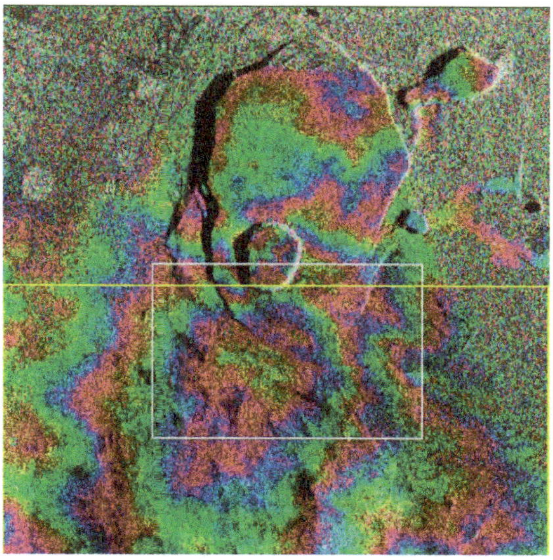

InSAR Image of Kilauea, Hawaii showing fringes caused by deformation of the terrain over a six-month period.

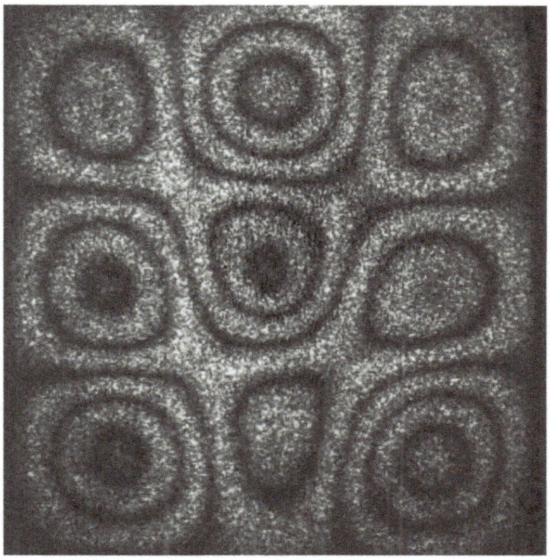

ESPI fringes showing a vibration mode of a clamped square plate

Interferometric synthetic aperture radar (InSAR) is a radar technique used in geodesy and remote sensing. Satellite synthetic aperture radar images of a geographic feature are taken on separate days, and changes that have taken place between radar images taken on the separate days are recorded as fringes similar to those obtained in holographic interferometry. The technique can monitor centimeter- to millimeter-scale deformation resulting from earthquakes, volcanoes and landslides, and also has uses in structural engineering, in particular for the monitoring of subsidence and structural stability. Fig 20 shows Kilauea, an active volcano in Hawaii. Data acquired using the space shuttle Endeavour's X-band Synthetic Aperture Radar on April 13, 1994 and October 4, 1994 were used to generate interferometric fringes, which were overlaid on the X-SAR image of Kilauea., also known as TV holography, uses video detection and recording to produce an image of the object upon which is superimposed a fringe pattern which represents the displacement of the object between recordings. The fringes are similar to those obtained in holographic interferometry.

When lasers were first invented, laser speckle was considered to be a severe drawback in using lasers to illuminate objects, particularly in holographic imaging because of the grainy image produced. It was later realized that speckle patterns could carry information about the object's surface deformations. Butters and Leendertz developed the technique of speckle pattern interferometry in 1970, and since then, speckle has been exploited in a variety of other applications. A photograph is made of the speckle pattern before deformation, and a second photograph is made of the speckle pattern after deformation. Digital subtraction of the two images results in a correlation fringe pattern, where the fringes represent lines of equal deformation. Short laser pulses in the nanosecond range can be used to capture very fast transient events. A phase problem exists: In the absence of other information, one cannot tell the difference between contour lines indicating a peak *versus* contour lines indicating a trough. To resolve the issue of phase ambiguity, ESPI may be combined with phase shifting methods.

A method of establishing precise geodetic baselines, invented by Yrjö Väisälä, exploited the low coherence length of white light. Initially, white light was split in two, with the reference beam "folded", bouncing back-and-forth six times between a mirror pair spaced precisely 1 m apart. Only if the test path was precisely 6 times the reference path would fringes be seen. Repeated applications of this procedure allowed precise measurement of distances up to 864 meters. Baselines thus established were used to calibrate geodetic distance measurement equipment, leading to a metrologically traceable scale for geodetic networks measured by these instruments. (This method has been superseded by GPS.)

Other uses of interferometers have been to study dispersion of materials, measurement of complex indices of refraction, and thermal properties. They are also used for three-dimensional motion mapping including mapping vibrational patterns of structures.

Actuators for Active Isolation

- Linear motors
- Pneumatic actuators
- Piezoelectric motors

Linear Motor

A linear motor is an electric motor that has had its stator and rotor "unrolled" so that instead of producing a torque (rotation) it produces a linear force along its length. However, linear motors are not necessarily straight. Characteristically, a linear motor's active section has ends, whereas more conventional motors are arranged as a continuous loop.

The most common mode of operation is as a Lorentz-type actuator, in which the applied force is linearly proportional to the current and the magnetic field $(\vec{F} = I\vec{L} \times \vec{B})$.

Many designs have been put forward for linear motors, falling into two major categories, low-acceleration and high-acceleration linear motors. Low-acceleration linear motors are suitable for maglev trains and other ground-based transportation applications. High-acceleration linear motors are normally rather short, and are designed to accelerate an object to a very high speed, for example see the coilgun.

Free-body diagram of a U-channel synchronous linear motor. The view is perpendicular to the channel axis. The two coils at centre are mechanically connected, and are energized in "quadrature" (with a phase difference of 90° (π/2 radians)). If the bottom coil (as shown) leads in phase, then the motor will move downward (in the drawing), and vice versa. (Not to scale)

High-acceleration linear motors are typically used in studies of hypervelocity collisions, as weapons, or as mass drivers for spacecraft propulsion. They are usually of the AC linear induction motor (LIM) design with an active three-phase winding on one side of the air-gap and a passive conductor plate on the other side. However, the direct current homopolar linear motor railgun is another high acceleration linear motor design. The low-acceleration, high speed and high power motors are usually of the linear synchronous motor (LSM) design, with an active winding on one side of the air-gap and an array of alternate-pole magnets on the other side. These magnets can be permanent magnets or energized magnets. The Shanghai Transrapid motor is an LSM.

Pneumatic Actuator

A pneumatic control valve actuator converts energy (typically in the form of compressed air) into mechanical motion. The motion can be rotary or linear, depending on the type of actuator.

Principle of Operation

A Pneumatic actuator mainly consists of a piston or a diaphragm which develops the motive power. It keeps the air in the upper portion of the cylinder, allowing air pressure to force the diaphragm or piston to move the valve stem or rotate the valve control element.

Globe control valve with pneumatic diaphragm actuator and "smart" positioner which will also feed back to the controller the actual valve position

Pneumatic rack and pinion actuators for valve controls of water pipes

Valves require little pressure to operate and usually double or *triple* the input force. The larger the size of the piston, the larger the output pressure can be. Having a larger piston can also be good if air supply is low, allowing the same forces with less input. These pressures are large enough to crush objects in the pipe. On 100 kPa input, you could lift a small car (upwards of 1,000 lbs) easily, and this is only a basic, small pneumatic valve. However, the resulting forces required of the stem would be too great and cause the valve stem to fail.

This pressure is transferred to the valve stem, which is connected to either the valve plug (see plug valve), butterfly valve etc. Larger forces are required in high pressure or high flow pipelines to allow the valve to overcome these forces, and allow it to move the valves moving parts to control the material flowing inside.

The valves input is the "control signal." This can come from a variety of measuring devices, and each different pressure is a different set point for a valve. A typical standard signal is 20–100 kPa. For example, a valve could be controlling the pressure in a vessel which has a constant out-flow, and a varied in-flow (varied by the actuator and valve). A pressure transmitter will monitor the pressure in the vessel and transmit a signal from 20–100 kPa. 20 kPa means there is no pressure, 100 kPa means there is full range pressure (can be varied by the transmitters calibration points). As the pressure rises in the vessel, the output of the transmitter rises, this increase in pressure is sent to the valve, which causes the valve to stroke downward, and start closing the valve, decreasing flow into the vessel, reducing the pressure in the vessel as excess pressure is evacuated through the out flow. This is called a direct acting process.

Types

Some types of pneumatic actuators include:

- Tie rod cylinders
- Rotary actuators
- Grippers
- Rodless actuators with magnetic linkage or rotary cylinders
- Rodless actuators with mechanical linkage
- Pneumatic artificial muscles
- Speciality actuators that combine rotary and linear motion—frequently used for clamping operations
- Vacuum generators

Piezoelectric Motor

A piezoelectric motor or piezo motor is a type of electric motor based upon the change in shape of a piezoelectric material when an electric field is applied. Piezoelectric motors make use of the converse piezoelectric effect whereby the material produces acoustic or ultrasonic vibrations in order to produce a linear or rotary motion. In one mechanism, the elongation in a single plane is used to make a series of stretches and position holds, similar to the way a caterpillar moves.

Insides of a slip-stick piezoelectric motor. Two piezoelectric crystals are visible that provide the mechanical torque.

Current Designs

Motors are made in both linear and rotary types.

One drive technique is to use piezoelectric ceramics to push a stator. These piezoelectric motors use three groups of crystals: two of which are Locking and one Motive, permanently connected to either the motor's casing or stator (not both) and sandwiched between the other two, which provides the motion. These piezoelectric motors are fundamentally stepping motors, with each step comprising either two or three actions, based on the locking type. These motors are also known as inchworm motors Another mechanism employs the use of surface acoustic waves (SAW) to generate linear or rotational motion.

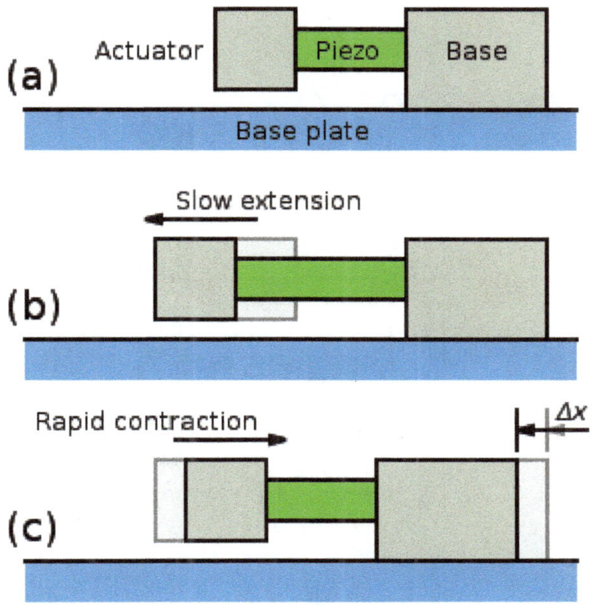

A slip-stick actuator.

A second drive technique is illustrated by the Squiggle motor, in which piezoelectric elements are bonded orthogonally to a nut and their ultrasonic vibrations rotate and translate a central lead screw. This is a direct drive mechanism.

Locking Mechanisms

The non-powered behavior of the first type of piezoelectric motor is one of two options: Normally Locked or Normally Free. When no power is being applied to a Normally Locked motor, the spindle or carriage (for rotary or linear types, respectively) will not move under external force. For a Normally Free motor, the spindle or carriage will move freely under external force. However, if both locking groups are powered at rest, a Normally Free motor will resist external force without providing any motive force.

A combination of mechanical latches and crystals could be used, but this would restrict the maximum stepping rate of the motor.

The non-power behavior of the second type of motor is locked, as the drive screw is locked by the threads on the nut. Thus it holds its position with the power off.

Stepping Actions

Regardless of locking type, stepping type piezoelectric motors — both linear and rotary — use the same mechanism to provide movement.

First, one group of *locking* crystals is activated — this gives one locked side and one unlocked side of the 'sandwich'.

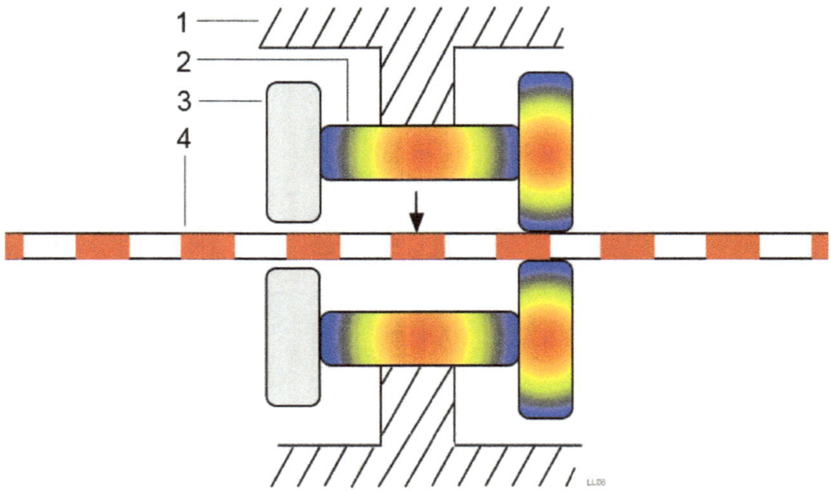

Piezoelectric "inchworm" motor

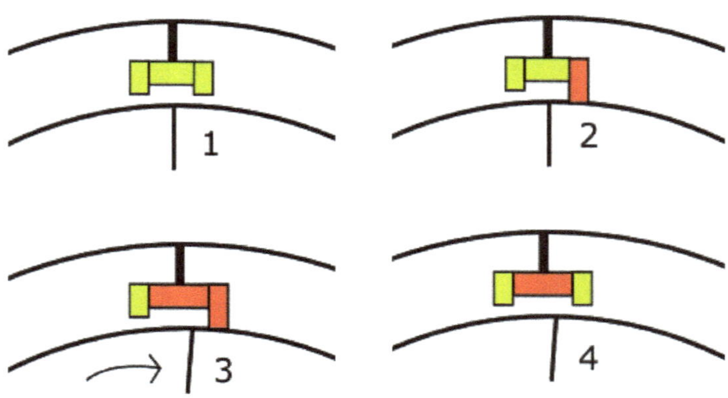

Stepping stages of **Normally Free** motor

Next, the *motive* crystal group is triggered and held — the expansion of this group moves the unlocked *locking* group along the motor path. This is the only stage where motor movement takes place.

Then the *locking* group triggered in stage one is released (in *Normally Locking* motors, the other is triggered). Then the *motive* group is released, retracting the 'trailing' *locking* group. Finally, both *locking* groups are returned to their default states.

Direct Drive Actions

The direct drive piezoelectric motor uses continuous ultrasonic vibrations to provide movement. A two-channel sinusoidal or square wave is applied to the piezoelectric elements at an ultrasonic frequency of 40 kHz to 200 kHz, matching the first bending resonant frequency of the threaded tube. This creates an orbital motion which drives the screw.

Speed and Precision

The growth and forming of piezoelectric crystals is a well-developed industry, yielding very uniform and consistent distortion for a given applied potential difference. This, combined with the minute scale of the distortions, gives the piezoelectric motor the ability to make very fine steps — manufacturers claim precision to the nanometer scale.

The high response rate and fast distortion of the crystals also allows the steps to be made at very high frequencies — upwards of 5 MHz. This gives a maximum linear speed of approximately 800 mm per second, or nearly 2.9 km/h.

A unique capability of piezoelectric motors is their ability to operate in the presence of strong magnetic fields, extending their applications to areas where traditional electromagnetic motors cannot be used - inside Nuclear Magnetic Resonance Antennas, for example. The maximum operating temperature is limited by the Curie temperature of the used piezoelectric ceramic and may exceed +250C.

Other Designs

Single Action

Piezo ratchet stepping motor.

Very simple single-action stepping motors can be made with piezoelectric crystals. For example, with a hard and rigid rotor-spindle coated with a thin layer of a softer material (like a polyurethane rubber), a series of angled piezoelectric transducers can be arranged.. When one group of transducers is triggered, the rotor will be pushed around one step. This design is not capable of such small or precise steps as more complex designs, but can reach higher speeds and are cheaper to manufacture.

REFERENCES

1. Reitherman, Robert (2012). Earthquakes and Engineers: An International History. Reston, VA: ASCE Press. ISBN 9780784410714.
2. Platus PhD, David L., SPIE International Society of Optical Engineering - July 1999, Optomechanical Engineering and Vibration Control Negative-Stiffness-Mechanism Vibration Isolation Systems
3. Harris, C., Piersol, A., Harris Shock and Vibration Handbook, Fifth Edition, McGraw-Hill, (2002), ISBN 0-07-137081-1
4. A.Kolesnikov Noise and vibration. Russia. Leningrad. Publ.Shipbuilding. 1988
5. Bunch, Bryan H; Hellemans, Alexander (April 2004). The History of Science and Technology. Houghton Mifflin Harcourt. p. 695. ISBN 978-0-618-22123-3.
6. Hariharan, P. (2007). Basics of Interferometry. Elsevier Inc. ISBN 0-12-373589-0.
7. Paschotta, Rüdiger. Optical Heterodyne Detection. RP Photonics Consulting GmbH. Retrieved 1 April 2012.
8. Poole, Ian. The superhet or superheterodyne radio receiver. Radio-Electronics. com. Retrieved 22 June 2012.
9. Mallick, S.; Malacara, D. (2007). Common-Path Interferometers. Optical Shop Testing. p. 97. doi:10.1002/9780470135976.ch3. ISBN 9780470135976.
10. Verma, R.K. (2008). Wave Optics. Discovery Publishing House. pp. 97–110. ISBN 81-8356-114-4.
11. Interferential Devices – Introduction. OPI – Optique pour l›Ingénieur. Retrieved 1 April 2012.
12. Ingram Taylor, Sir Geoffrey (1909). Interference Fringes with Feeble Light (PDF). Proc. Cam. Phil. Soc. 15: 114. Retrieved 2 January 2013.
13. Jönsson, C (1961). Elektroneninterferenzen an mehreren künstlich hergestellten Feinspalten. Zeitschrift für Physik. 161 (4): 454–474. Bibcode:1961ZPhy..161..454J. doi:10.1007/BF01342460.
14. Jönsson, C (1974). Electron diffraction at multiple slits. American Journal of Physics. 4: 4–11. Bibcode:1974AmJPh..42....4J. doi:10.1119/1.1987592.
15. Arndt, M.; Zeilinger, A. (2004). Heisenberg›s Uncertainty and Matter Wave Interferometry with Large Molecules. In Buschhorn, G. W.; Wess, J. Fundamental Physics-- Heisenberg and Beyond: Werner Heisenberg Centennial Symposium Developments in Modern Physics. Springer. pp. 35–52. ISBN 3540202013. Retrieved 26 May 2012.

16. Carroll, Brett. Simple Lloyd›s Mirror (PDF). American Association of Physics Teachers. Retrieved 5 April 2012.

17. Serway, R.A.; Jewett, J.W. (2010). Principles of physics: a calculus-based text, Volume 1. Brooks Cole. pp. 905–905. ISBN 0-534-49143-X.

18. Nolte, David D. (2012). Optical Interferometry for Biology and Medicine. Springer. pp. 17–26. ISBN 1-4614-0889-X.

19. Guideline for Use of Fizeau Interferometer in Optical Testing (PDF). NASA. Retrieved 8 April 2012.

20. Interferential devices – Fizeau Interferometer. Optique pour l›Ingénieur. Retrieved 8 April 2012.

21. Zetie, K.P.; Adams, S.F.; Tocknell, R.M. How does a Mach–Zehnder interferometer work? (PDF). Physics Department, Westminster School, London. Retrieved 8 April 2012.

22. Ashkenas, Harry I. (1950). The design and construction of a Mach-Zehnder interferometer for use with the GALCIT Transonic Wind Tunnel. Engineer›s thesis. California Institute of Technology.

23. Betzler, Klaus. Fabry-Perot Interferometer (PDF). Fachbereich Physik, Universität Osnabrück. Retrieved 8 April 2012.

24. Michelson, A.A.; Morley, E.W. (1887). On the Relative Motion of the Earth and the Luminiferous Ether (PDF). American Journal of Science. 34 (203): 333–345. doi:10.2475/ajs.s3-34.203.333.

25. Miller, Dayton C. (1933). The Ether-Drift Experiment and the Determination of the Absolute Motion of the Earth. Reviews of Modern Physics. 5 (3): 203–242. Bibcode:1933RvMP....5..203M. doi:10.1103/RevModPhys.5.203. White light fringes were chosen for the observations because they consist of a small group of fringes having a central, sharply defined black fringe which forms a permanent zero reference mark for all readings.

26. Müller, H.; Herrmann, S.; Braxmaier, C.; Schiller, S.; Peters, A. (2003). Modern Michelson-Morley experiment using cryogenic optical resonators. Phys. Rev. Lett. 91 (2): 020401. arXiv:physics/0305117Freely accessible. Bibcode:2003PhRvL..91b0401M. doi:10.1103/PhysRevLett.91.020401. PMID 12906465.

27. Eisele, C.; Nevsky, A.; Schiller, S. (2009). Laboratory Test of the Isotropy of Light Propagation at the 10-17 Level. Physical Review Letters. 103 (9): 090401. Bibcode:2009PhRvL.103i0401E. doi:10.1103/PhysRevLett.103.090401. PMID 19792767.

28. Herrmann, S.; Senger, A.; Möhle, K.; Nagel, M.; Kovalchuk, E.; Peters, A. (2009). Rotating optical cavity experiment testing Lorentz invariance at the 10-17 level. Physical Review D. 80 (10). arXiv:1002.1284Freely accessible. Bibcode:2009PhRvD..80j5011H. doi:10.1103/PhysRevD.80.105011.

29. Scherrer, P.H.; Bogart, R.S.; Bush, R.I.; Hoeksema, J.; Kosovichev, A.G.; Schou, J. (1995). The Solar Oscillations Investigation – Michelson Doppler Imager. Solar

Physics. 162: 129–188. Bibcode:1995SoPh..162..129S. doi:10.1007/BF00733429. Retrieved 2 April 2012.

30. Stroke, G.W.; Funkhouser, A.T. (1965). Fourier-transform spectroscopy using holographic imaging without computing and with stationary interferometers (PDF). Physics Letters. 16 (3): 272–274. Bibcode:1965PhL....16..272S. doi:10.1016/0031-9163(65)90846-2. Retrieved 2 April 2012.
31. Gary, G.A.; Balasubramaniam, K.S. Additional Notes Concerning the Selection of a Multiple-Etalon System for ATST (PDF). Advanced Technology Solar Telescope. Retrieved 29 April 2012.
32. Spectrometry by Fourier transform. OPI – Optique pour l›Ingénieur. Retrieved 3 April 2012.
33. Halloween 2003 Solar Storms: SOHO/EIT Ultraviolet, 195 Ã". NASA/Goddard Space Flight Center Scientific Visualization Studio. Retrieved 20 June 2012.
34. LIGO -Laser Interferometer Gravitational-Wave Observatory. Caltech/MIT. Retrieved 4 April 2012.
35. Castelvecchi, Davide; Witze, Alexandra (11 February 2016). Einstein›s gravitational waves found at last. Nature News. doi:10.1038/nature.2016.19361. Retrieved 11 February 2016.
36. Chevalerias, R.; Latron, Y.; Veret, C. (1957). Methods of Interferometry Applied to the Visualization of Flows in Wind Tunnels. Journal of the Optical Society of America. 47 (8): 703. doi:10.1364/JOSA.47.000703.
37. Ristić, Slavica. Flow visualization techniques in wind tunnels – optical methods (Part II) (PDF). Military Technical Institute, Serbia. Retrieved 6 April 2012.
38. Paris, M.G.A. (1999). Entanglement and visibility at the output of a Mach-Zehnder interferometer (PDF). Physical Review A. 59 (2): 1615–1621. arXiv:quant-ph/9811078Freely accessible. Bibcode:1999PhRvA..59.1615P. doi:10.1103/PhysRevA.59.1615. Retrieved 2 April 2012.
39. Haack, G. R.; Förster, H.; Büttiker, M. (2010). Parity detection and entanglement with a Mach-Zehnder interferometer. Physical Review B. 82 (15). arXiv:1005.3976Freely accessible. Bibcode:2010PhRvB..82o5303H. doi:10.1103/PhysRevB.82.155303.
40. Monnier, John D (2003). Optical interferometry in astronomy (PDF). Reports on Progress in Physics. 66 (5): 789–857. arXiv:astro-ph/0307036Freely accessible. Bibcode:2003RPPh...66..789M. doi:10.1088/0034-4885/66/5/203.
41. Cosmic Calibration. www.eso.org. Retrieved 10 October 2016.
42. Malbet, F.; Kern, P.; Schanen-Duport, I.; Berger, J.-P.; Rousselet-Perraut, K.; Benech, P. (1999). Integrated optics for astronomical interferometry. Astron. Astrophys. Suppl. Ser. 138: 1–10. arXiv:astro-ph/9907031Freely accessible. Bibcode:1999A&AS..138....1D. doi:10.1051/aas:1999263.
43. Baldwin, J.E.; Haniff, C.A. (2002). The application of interferometry to optical astronomical imaging (PDF). Phil. Trans. R. Soc. Lond. A. 360 (1794): 969–986. Bibcode:2002RSPTA.360..969B. doi:10.1098/rsta.2001.0977. Retrieved 10 April

2012.

44. Zhao, M.; Gies, D.; Monnier, J. D.; Thureau, N.; Pedretti, E.; Baron, F.; Merand, A.; Ten Brummelaar, T.; McAlister, H.; Ridgway, S. T.; Turner, N.; Sturmann, J.; Sturmann, L.; Farrington, C.; Goldfinger, P. J. (2008). First Resolved Images of the Eclipsing and Interacting Binary β Lyrae. The Astrophysical Journal. 684 (2): L95. arXiv:0808.0932Freely accessible. Bibcode:2008ApJ...684L..95Z. doi:10.1086/592146.

45. Gerlich, S.; Eibenberger, S.; Tomandl, M.; Nimmrichter, S.; Hornberger, K.; Fagan, P. J.; Tüxen, J.; Mayor, M.; Arndt, M. (2011). Quantum interference of large organic molecules. Nature Communications. 2: 263–. Bibcode:2011NatCo...2E.263G. doi:10.1038/ncomms1263. PMC 3104521Freely accessible. PMID 21468015.

46. Hornberger, Klaus; Gerlich, Stefan; Haslinger, Philipp; Nimmrichter, Stefan; Arndt, Markus (2012-02-08). \textit{Colloquium} : Quantum interference of clusters and molecules. Reviews of Modern Physics. 84 (1): 157–173. Bibcode:2012RvMP...84..157H. doi:10.1103/RevModPhys.84.157.

47. Eibenberger, Sandra; Gerlich, Stefan; Arndt, Markus; Mayor, Marcel; Tüxen, Jens (2013-08-14). Matter–wave interference of particles selected from a molecular library with masses exceeding 10000 amu. Physical Chemistry Chemical Physics. 15 (35): 14696–700. Bibcode:2013PCCP...1514696E. doi:10.1039/C3CP51500A. ISSN 1463-9084. PMID 23900710.

48. Lehmann, M; Lichte, H (December 2002). Tutorial on off-axis electron holography. Microsc. Microanal. 8 (6): 447–66. Bibcode:2002MiMic...8..447L. doi:10.1017/S1431927602029938. PMID 12533207.

49. Tonomura, A. (1999). Electron Holography, 2nd ed. Springer. ISBN 3-540-64555-1.

50. Klein, T. (2009). Neutron interferometry: A tale of three continents. Europhysics News. 40 (6): 24–20. Bibcode:2009ENews..40...24K. doi:10.1051/epn/2009802.

51. Dimopoulos, S.; Graham, P.W.; Hogan, J.M.; Kasevich, M.A. (2008). General Relativistic Effects in Atom Interferometry. Phys. Rev. D. 78 (042003). arXiv:0802.4098Freely accessible. Bibcode:2008PhRvD..78d2003D. doi:10.1103/PhysRevD.78.042003.

52. Mariani, Z.; Strong, K.; Wolff, M.; et al. (2012). Infrared measurements in the Arctic using two Atmospheric Emitted Radiance Interferometers. Atmos. Meas. Tech. 5: 329–344. Bibcode:2012AMT.....5..329M. doi:10.5194/amt-5-329-2012.

53. Mantravadi, M. V.; Malacara, D. (2007). Newton, Fizeau, and Haidinger Interferometers. Optical Shop Testing. p. 1. doi:10.1002/9780470135976.ch1. ISBN 9780470135976.

54. Malacara, D. (2007). Twyman–Green Interferometer. Optical Shop Testing. p. 46. doi:10.1002/9780470135976.ch2. ISBN 9780470135976.

55. Michelson, A. A. (1918). On the Correction of Optical Surfaces. Proceedings of the National Academy of Sciences of the United States of America. 4 (7): 210–212. Bibcode:1918PNAS....4..210M. doi:10.1073/pnas.4.7.210. PMC 1091444Freely accessible. PMID 16576300.

56. Interferential Devices – Twyman-Green Interferometer. OPI – Optique pour l›Ingénieur. Retrieved 4 April 2012.
57. Heideman, R. G.; Kooyman, R. P. H.; Greve, J. (1993). Performance of a highly sensitive optical waveguide Mach-Zehnder interferometer immunosensor. Sensors and Actuators B: Chemical. 10 (3): 209–217. doi:10.1016/0925-4005(93)87008-D.
58. Oliver, W. D.; Yu, Y.; Lee, J. C.; Berggren, K. K.; Levitov, L. S.; Orlando, T. P. (2005). Mach-Zehnder Interferometry in a Strongly Driven Superconducting Qubit. Science. 310 (5754): 1653–1657. arXiv:cond-mat/0512691Freely accessible. Bibcode:2005Sci...310.1653O. doi:10.1126/science.1119678. PMID 16282527.
59. Nieradko, Ł.; Gorecki, C.; JóZwik, M.; Sabac, A.; Hoffmann, R.; Bertz, A. (2006). Fabrication and optical packaging of an integrated Mach-Zehnder interferometer on top of a movable micromirror. Journal of Microlithography, Microfabrication, and Microsystems. 5 (2): 023009. Bibcode:2006JMM&M...5b3009N. doi:10.1117/1.2203366.
60. Burge, J. H.; Zhao, C.; Dubin, M. (2010). Measurement of aspheric mirror segments using Fizeau interferometry with CGH correction (PDF). Proceedings of SPIE. 7739: 773902. doi:10.1117/12.857816.
61. Anderson, R.; Bilger, H.R.; Stedman, G.E. (1994). Sagnac effect A century of Earth-rotated interferometers (PDF). Am. J. Phys. 62 (11): 975–985. Bibcode:1994AmJPh..62..975A. doi:10.1119/1.17656. Retrieved 30 March 2012.
62. Golio, Mike (2007). RF and Microwave Applications and Systems. CRC Press. pp. 14.1–14.17. ISBN 0849372194. Retrieved 27 June 2012.
63. Paschotta, Rüdiger. Self-heterodyne Linewidth Measurement. RP Photonics. Retrieved 22 June 2012.
64. Optical Frequency Comb. National Research Council, Canada. Retrieved 23 June 2012.
65. Paschotta, Rüdiger. Frequency Combs. RP Photonics. Retrieved 23 June 2012.
66. Schmit, J. (1993). Spatial and temporal phase-measurement techniques: a comparison of major error sources in one dimension. Proceedings of SPIE. 1755. pp. 202–201. doi:10.1117/12.140770.
67. Larkin, K.G. (1996). Efficient nonlinear algorithm for envelope detection in white light interferometry (PDF). Journal of the Optical Society of America. 13 (4): 832–843. Bibcode:1996JOSAA..13..832L. doi:10.1364/JOSAA.13.000832. Retrieved 1 April 2012.
68. ISO. (2013). 25178-604:2013(E): Geometrical product specification (GPS) – Surface texture: Areal – Nominal characteristics of non-contact (coherence scanning interferometric microscopy) instruments (2013(E) ed.). Geneva: International Organization for Standardization.
69. Harasaki, A.; Schmit, J.; Wyant, J. C. (2000). Improved vertical-scanning interferometry (PDF). Applied Optics. 39 (13): 2107–2115. Bibcode:2000ApOpt..39.2107H. doi:10.1364/AO.39.002107. Retrieved 21 May 2012.

70. De Groot, P (2015). Principles of interference microscopy for the measurement of surface topography. Advances in Optics and Photonics. 7: 1–65. doi:10.1364/AOP.7.000001.

71. Olszak, A.G.; Schmit, J.; Heaton, M.G. Interferometry: Technology and Applications (PDF). Bruker. Retrieved 1 April 2012.

72. de Groot, Peter; Deck, Leslie (1995). Surface Profiling by Analysis of White-light Interferograms in the Spatial Frequency Domain. Journal of Modern Optics. 42 (2): 389–401. Bibcode:1995JMOp...42..389D. doi:10.1080/09500349514550341.

73. Phase-Shifting Interferometry for Determining Optical Surface Quality. Newport Corporation. Retrieved 12 May 2012.

74. How Phase Interferometers work. Graham Optical Systems. 2011. Retrieved 12 May 2012.

75. Schreiber, H.; Bruning, J. H. (2007). Phase Shifting Interferometry. Optical Shop Testing. p. 547. doi:10.1002/9780470135976.ch14. ISBN 9780470135976.

76. Sommargren, G. E. (1986). US Patent 4,594,003.

77. Ferraro, P.; Paturzo, M ; Grilli, S. (2007). Optical wavefront measurement using a novel phase-shifting point-diffraction interferometer. SPIE. Retrieved 26 May 2012.

78. P. de Groot, J., Interference Microscopy for Surface Structure Analysis, in Handbook of Optical Metrology, edited by T. Yoshizawa, chapt.31, pp. 791-828, (CRC Press, 2015).

79. Schmit, J.; Creath, K.; Wyant, J. C. (2007). Surface Profilers, Multiple Wavelength, and White Light Intereferometry. Optical Shop Testing. p. 667. doi:10.1002/9780470135976.ch15. ISBN 9780470135976.

80. HDVSI – Introducing High Definition Vertical Scanning Interferometry for Nanotechnology Research from Veeco Instruments. Veeco. Retrieved 21 May 2012.

81. Plucinski, J.; Hypszer, R.; Wierzba, P.; Strakowski, M.; Jedrzejewska-Szczerska, M.; Maciejewski, M.; Kosmowski, B.B. (2008). Optical low-coherence interferometry for selected technical applications (PDF). Bulletin of the Polish Academy of Sciences. 56 (2): 155–172. Retrieved 8 April 2012.

82. Yang, C.-H.; Wax, A; Dasari, R.R.; Feld, M.S. (2002). 2π ambiguity-free optical distance measurement with subnanometer precision with a novel phase-crossing low-coherence interferometer (PDF). Optics Letters. 27 (2): 77–79. Bibcode:2002OptL...27...77Y. doi:10.1364/OL.27.000077.

83. Hitzenberger, C. K.; Sticker, M.; Leitgeb, R.; Fercher, A. F. (2001). Differential phase measurements in low-coherence interferometry without 2pi ambiguity. Optics Letters. 26 (23): 1864–1866. Bibcode:2001OptL...26.1864H. doi:10.1364/ol.26.001864. PMID 18059719.

84. Wojtek J. Walecki, Kevin Lai, Vitalij Souchkov, Phuc Van, SH Lau, Ann Koo physica status solidi (c) Volume 2, Issue 3, Pages 984–989

85. W. J. Walecki et al. Non-contact fast wafer metrology for ultra-thin patterned

wafers mounted on grinding and dicing tapes Electronics Manufacturing Technology Symposium, 2004. IEEE/CPMT/SEMI 29th International Volume, Issue, July 14–16, 2004 Page(s): 323–325

86. Coating Thickness Measurement. Lumetrics, Inc. Retrieved 28 October 2013.
87. Typical profilometry measurements. Novacam Technologies, Inc. Retrieved 25 June 2012.
88. Holographic interferometry. Oquagen. 2008. Retrieved 22 May 2012.
89. Hecht, Jeff (1998). Laser, Light of a Million Uses. Dover Publications, Inc. pp. 229–230. ISBN 0-486-40193-6.
90. Fein, H (September 1997). Holographic Interferometry: Nondestructive tool (PDF). The Industrial Physicist: 37–39.
91. PIA01762: Space Radar Image of Kilauea, Hawaii. NASA/JPL. 1999. Retrieved 17 June 2012.
92. Jones R & Wykes C, Holographic and Speckle Interferometry, 1989, Cambridge University Press
93. Butters, J. N.; Leendertz, J. A. (1971). A double exposure technique for speckle pattern interferometry. Journal of Physics E: Scientific Instruments. 4 (4): 277–279. Bibcode:1971JPhE....4..277B. doi:10.1088/0022-3735/4/4/004.
94. Dvořáková, P.; Bajgar, V.; Trnka, J. (2007). Dynamic Electronic Speckle Pattern Interferometry in Application to Measure Out-Of-Plane Displacement (PDF). Engineering Mechanics. 14 (1/2): 37–44.
95. Moustafa, N. A.; Hendawi, N. (2003). Comparative Phase-Shifting Digital Speckle Pattern Interferometry Using Single Reference Beam Technique (PDF). Egypt. J. Sol. 26 (2): 225–229. Retrieved 22 May 2012.
96. Buga, A.; Jokela, J.; Putrimas, R. Traceability, stability and use of the Kyviskes calibration baseline–the first 10 years. Environmental Engineering, The 7th International Conference (PDF). Vilnius Gediminas Technical University. pp. 1274–1280. Retrieved 9 April 2012.
97. Nolte, David D. (2012). Optical Interferometry for Biology and Medicine. Springer. ISBN 1-4614-0889-X.
98. Huang, D.; Swanson, E.A.; Lin, C.P.; Schuman, J.S.; Stinson, W.G.; Chang, W.; Hee, M.R.; Flotte, T.; Gregory, K.; Puliafito, C.A.; Fujimoto, J.G. (1991). Optical Coherence Tomography (PDF). Science. 254 (5035): 1178–81. Bibcode:1991Sci...254.1178H. doi:10.1126/science.1957169. PMC 4638169Freely accessible. PMID 1957169. Retrieved 10 April 2012.
99. Fercher, A.F. (1996). Optical Coherence Tomography (PDF). Journal of Biomedical Optics. 1 (2): 157–173. Bibcode:1996JBO.....1..157F. doi:10.1117/12.231361. Retrieved 10 April 2012.
100. Lang, Walter. Nomarski Differential Interference-Contrast Microscopy (PDF). Carl Zeiss, Oberkochen. Retrieved 10 April 2012.
101. Wax, A.; Pyhtila, J. W.; Graf, R. N.; Nines, R.; Boone, C. W.; Dasari, R. R.; Feld, M. S.; Steele, V. E.; Stoner, G. D. (2005). Prospective grading of neoplastic change

in rat esophagus epithelium using angle-resolved low-coherence interferometry. Journal of Biomedical Optics. 10 (5): 051604. Bibcode:2005JBO....10e1604W. doi:10.1117/1.2102767. PMID 16292952.

102. Pyhtila, J. W.; Chalut, K. J.; Boyer, J. D.; Keener, J.; d›Amico, T.; Gottfried, M.; Gress, F.; Wax, A. (2007). In situ detection of nuclear atypia in Barrett›s esophagus by using angle-resolved low-coherence interferometry. Gastrointestinal Endoscopy. 65 (3): 487–491. doi:10.1016/j.gie.2006.10.016. PMID 17321252.
103. Fitzgerald, Richard (2000). Phase-sensitive x-ray imaging. Physics Today. 53 (7): 23. Bibcode:2000PhT....53g..23F. doi:10.1063/1.1292471.
104. David, C, Nohammer, B, Solak, H H, & Ziegler E (2002). Differential x-ray phase contrast imaging using a shearing interferometer. Applied Physics Letters. 81 (17): 3287–3289. Bibcode:2002ApPhL..81.3287D. doi:10.1063/1.1516611.
105. Wilkins, S W, Gureyev, T E, Gao, D, Pogany, A & Stevenson, A W (1996). Phase-contrast imaging using polychromatic hard X-rays. Nature. 384 (6607): 335–338. Bibcode:1996Natur.384..335W. doi:10.1038/384335a0.
106. Miao, Houxun; Panna, Alireza; Gomella, Andrew A.; Bennett, Eric E.; Znati, Sami; Chen, Lei; Wen, Han. A universal moiré effect and application in X-ray phase-contrast imaging. Nature Physics. doi:10.1038/nphys3734.
107. Davis, T J, Gao, D, Gureyev, T E, Stevenson, A W & Wilkins, S W (1995). Phase-contrast imaging of weakly absorbing materials using hard X-rays. Nature. 373 (6515): 595–598. Bibcode:1995Natur.373..595D. doi:10.1038/373595a0.
108. Momose, A, Takeda, T, Itai, Y & Hirano, K (1996). Phase-contrast X-ray computed tomography for observing biological soft tissues. Nature Medicine. 2 (4): 473–475. doi:10.1038/nm0496-473. PMID 8597962.

CHAPTER 7

Earthquake Resistant Structures

Chapter Contents

Experience in past earthquakes has demonstrated that many common buildings and typical methods of construction lack basic resistance to earthquake forces. In most cases this resistance can be achieved by following simple, inexpensive principles of good building construction practice.

There are a wide variety of earthquake effects – these might include a chasm opening up or a drop of many meters across a fault line.

Therefore, it is not possible to design an earthquake proof building which is guaranteed to resist all possible earthquakes.

However, it is possible during your design and construction process to build in a number of earthquake resistant features by applying earthquake engineering techniques, which would increase enormously the chances of survival of both buildings and their occupants.

INTRODUCTION

Earthquake-resistant structures are structures designed to withstand earthquakes. While no structure can be entirely immune to damage from earthquakes, the goal of earthquake-resistant construction is to erect structures that fare better during seismic activity than their conventional counterparts.

According to building codes, earthquake-resistant structures are intended to withstand the largest earthquake of a certain probability that is likely to occur at their location. This means the loss of life should be minimized by preventing collapse of the buildings for rare earthquakes while the loss of functionality should be limited for more frequent ones.

To combat earthquake destruction, the only method available to ancient architects was to build their landmark structures to last, often by making them excessively stiff and strong, like the El Castillo pyramid at Chichen Itza.

Earthquake-proof and massive pyramid El Castillo, Chichen Itza.

Currently, there are several design philosophies in earthquake engineering, making use of experimental results, computer simulations and observations from past earthquakes to offer the required performance for the seismic threat at the site of interest. These range from appropriately sizing the structure to be strong and ductile enough to survive the shaking with an acceptable damage, to equipping it with base isolation or using structural vibration control technologies to minimize any forces and deformations. While the former is the method typically applied in most earthquake-resistant structures, important facilities, landmarks and cultural heritage buildings use the more advanced (and expensive) techniques of isolation or control to survive strong shaking with minimal damage. Examples of such applications are the Cathedral of Our Lady of the Angels and the Acropolis Museum.

TRENDS AND PROJECTS

Some of the new trends and/or projects in the field of earthquake engineering structures are presented.

Building Materials

Based on experience in earthquakes in Eastern European and in Central Asian countries where precast concrete has been widely used as construction material, it can be concluded that their seismic performance has been fairly satisfactory. Based on studies in New Zealand, relating to Christchurch earthquakes, precast concrete designed and installed in accordance with modern codes performed well, as did houses and structures made of timber According to the Earthquake Engineering Research Institute, precast panel buildings had good durability during the earthquake in Armenia, compared to precast frame-panels.

Earthquake Shelter

One Japanese construction company has developed a six-foot cubical shelter, presented as an alternative to earthquake-proofing an entire building.

Concurrent Shake-Table Testing

Concurrent shake-table testing of two or more building models is a vivid, persuasive and effective way to validate earthquake engineering solutions experimentally.

Thus, two wooden houses built before adoption of the 1981 Japanese Building Code were moved to E-Defense for testing (see both pictures aside). The left house was reinforced to enhance its seismic resistance, while the other one was not. These two models were set on E-Defense platform and tested simultaneously .

Combined Vibration control Solution

Close-up of abutment of seismically retrofitted Municipal Services Building in Glendale, CA

Seismically retrofitted Municipal Services Building in Glendale, CA

Designed by architect Merrill W. Baird of Glendale, working in collaboration with A. C. Martin Architects of Los Angeles, the Municipal Services Building at 633 East Broadway, Glendale was completed in 1966 . Prominently sited at the corner of East Broadway and Glendale Avenue, this civic building serves as a heraldic element of Glendale's civic center.

In October 2004 Architectural Resources Group (ARG) was contracted by Nabih Youssef & Associates, Structural Engineers, to provide services regarding a historic resource assessment of the building due to a proposed seismic retrofit.

In 2008, the Municipal Services Building of the City of Glendale, California was seismically retrofitted using an innovative combined vibration control solution: the existing elevated building foundation of the building was put on high damping rubber bearings.

Steel Plate Walls System

A steel plate shear wall (SPSW) consists of steel infill plates bounded by a column-beam system. When such infill plates occupy each level within a framed bay of a structure, they constitute a SPSW system. Whereas most earthquake resistant construction methods are adapted from older systems, SPSW was invented entirely to withstand seismic activity.

Coupled steel plate shear walls, Seattle

The Ritz-Carlton/JW Marriott hotel building engaging the advanced steel plate shear walls system, LA

SPSW behavior is analogous to a vertical plate girder cantilevered from its base. Similar to plate girders, the SPSW system optimizes component performance by taking advantage of the post-buckling behavior of the steel infill panels.

The Ritz-Carlton/JW Marriott hotel building, a part of the LA Live development in Los Angeles, California, is the first building in Los Angeles that uses an advanced steel plate shear wall system to resist the lateral loads of strong earthquakes and winds.

Kashiwazaki-Kariwa Nuclear Power Plant is partially upgraded

The Kashiwazaki-Kariwa Nuclear Power Plant, the largest nuclear generating station in the world by net electrical power rating, happened to be near the epicenter of the strongest M_w 6.6 July 2007 Chūetsu offshore earthquake. This initiated an extended shutdown for structural inspection which indicated that a greater earthquake-proofing was needed before operation could be resumed.

On May 9, 2009, one unit (Unit 7) was restarted, after the seismic upgrades. The test run had to continue for 50 days. The plant had been completely shut down for almost 22 months following the earthquake.

Seismic Test of Seven-Story Building

A destructive earthquake struck a lone, wooden condominium in Japan . The experiment was webcast live on July 14, 2009 to yield insight on how to make wooden structures stronger and better able to withstand major earthquakes .

The Miki shake at the Hyogo Earthquake Engineering Research Center is the capstone experiment of the four-year NEESWood project, which receives its primary support from the U.S. National Science Foundation Network for Earthquake Engineering Simulation (NEES) Program.

"NEESWood aims to develop a new seismic design philosophy that will provide the necessary mechanisms to safely increase the height of wood-frame structures in active seismic zones of the United States, as well as mitigate earthquake damage to low-rise wood-frame structures," said Rosowsky, Department of Civil Engineering at Texas A&M University. This philosophy is based on the application of seismic damping systems for wooden buildings. The systems, which can be installed inside the walls of most wooden buildings, include strong metal frame, bracing and dampers filled with viscous fluid.

Superframe RC Earthquake proof structure

The proposed system is composed of core walls, hat beams incorporated into the top level, outer columns and viscous dampers vertically installed between the tips of the hat beams and the outer columns. During an earthquake, the hat beams and outer columns act as outriggers and reduce the overturning moment in the core, and the installed dampers also reduce the moment and the lateral deflection of the structure. This innovative system can eliminate inner beams and inner columns on each floor, and thereby provide buildings with column-free floor space even in highly seismic regions.

TSUNAMI-PROOF BUILDING

A tsunami-proof building is a purposefully designed building which will, through its design integrity, withstand and survive the forces of a tsunami wave or extreme storm surge. It is hydrodynamically shaped to offer protection from high waves.

An example of such an architecture is where a laminar flow around a building will protect the walls. The structure can also rest on a hollow masonry block that for example can hold a body of water to sustain a family. A design can include battered walls, cantilever steps and a wooden superstructure with the walls jutting out. Bamboo ply panels can be added to cover the sides.

A structure like this, concomitant with its mechanical strength, will provide its occupants with independent potable water storage for an extended period of time. The first example known has been constructed at Poovar Island in southern Kerala, India.

REFERENCES

1. Seismology Committee (1999). Recommended Lateral Force Requirements and Commentary. Structural Engineers Association of California.
2. Civil Engineering Dictionary: PreCast Concrete in Earthquake Regions
3. Precast New Zealand Inc: Precast concrete and seismic issues
4. Earthquake Resistant Homes

5. Earthquake Engineering Research Institute: Precast concrete panel building damage, comparing the performance of precast frame-panel
6. Kharrazi, M.H.K., 2005, "Rational Method for Analysis and Design of Steel Plate Walls," Ph.D. Dissertation, University of British Columbia, Vancouver, Canada,
7. Reitherman, Robert (2012). Earthquakes and Engineers: An International History. Reston, VA: ASCE Press. pp. 356–357. ISBN 9780784410714.
8. "Profits shaken at Tepco". World Nuclear News. 31 July 2007. Retrieved 2007-08-01.
9. Asahi.com. Quake exposes nuke-plant danger. July 18, 2007.
10. A Survey on concepts of design and executing of Superframe RC Earthquake proof Structures" (2016) by Kiarash Khodabakhshi ISBN 9783668208704
11. Standing tall against tsunami

CHAPTER 8

Disaster Management

Chapter Contents

Emergency management is the discipline of dealing with and avoiding risks, particularly those that have catastrophic consequences for communities, regions, or entire countries. It is the dynamic process of preparing for, mitigating, responding to and recovering from an emergency. Planning, though critical, is not the only component. Training, conducting drills, testing equipment and coordinating activities with the community are other important functions. Effective emergency management relies on the integration of emergency plans at all levels of government and non-government, including individuals and community organizations.

INTRODUCTION

Disaster management (or emergency management) is the creation of plans through which communities reduce vulnerability to hazards and cope with disasters. Disaster management does not avert or eliminate the threats; instead, it focuses on creating plans to decrease the effect of disasters. Failure to create a plan could lead to human mortality, lost revenue, and damage to assets. Currently in the United States 60 percent of businesses do not have emergency management plans. Events covered by disaster management include acts of terrorism, industrial sabotage, fire, natural disasters (such as earthquakes, hurricanes, etc.), public disorder, industrial accidents, and communication failures.

EMERGENCY PLANNING IDEALS

If possible, emergency planning should aim to prevent emergencies from occurring, and failing that, should develop a good action plan to mitigate the results and effects of any emergencies. As time goes on, and more data becomes available, usually through the study of emergencies as they occur, a plan should evolve. The development of emergency plans is a cyclical process, common to many risk management disciplines, such as Business Continuity and Security Risk Management, as set out below:

- Recognition or identification of risks
- Ranking or evaluation of risks
    - Responding to significant risks
    - Tolerate
    - Treat
    - Transfer
    - Terminate
- Resourcing controls
- Reaction Planning
- Reporting & monitoring risk performance
- Reviewing the Risk Management framework

There are a number of guidelines and publications regarding Emergency Planning, published by various professional organizations such as ASIS, National Fire Protection Association (NFPA), and the International Association of Emergency Managers (IAEM). There are very few Emergency Management specific standards, and emergency management as a discipline tends to fall under business resilience standards.

In order to avoid, or reduce significant losses to a business, emergency managers should work to identify and anticipate potential risks, hopefully to reduce their probability of occurring. In the event that an emergency does occur, managers should have a plan prepared to mitigate the effects of that emergency, as well as to ensure Business Continuity of critical operations post-incident. It is essential for an organization to include procedures for determining whether an emergency situation has occurred and at what point an emergency management plan should be activated.

IMPLEMENTATION IDEALS

An emergency plan must be regularly maintained, in a structured and methodical manner, to ensure it is up-to-date in the event of an emergency. Emergency managers generally follow a common process to anticipate, assess, prevent, prepare, respond and recover from an incident.

Pre-Incident Training and Testing

Emergency management plans and procedures should include the identification of appropriately trained staff members responsible for decision-making when an emergency occurs. Training plans should include internal people, contractors and civil protection partners, and should state the nature and frequency of training and testing.

A team of emergency responders performs a training scenario involving anthrax.

Testing of a plan's effectiveness should occur regularly. In instances where several business or organizations occupy the same space, joint emergency plans, formally agreed to by all parties, should be put into place.

Communicating and Incident Assessment

Communication is one of the key issues during any emergency, pre-planning of communications is critical. Miscommunication can easily result in emergency events escalating unnecessarily. Once an emergency has been identified a comprehensive assessment evaluating the level of impact and its financial implications should be undertaken. Following assessment, the appropriate plan or response to be activated will depend on a specific pre-set criteria within the emergency plan. The steps necessary should be prioritized to ensure critical functions are operational as soon as possible.

PHASES AND PERSONAL ACTIVITIES

Emergency management consists of five phases: prevention, mitigation, preparedness, response and recovery. http://www.fema.gov/mission-areas

Prevention

It focuses on preventing the human hazard, primarily from potential natural disasters or terrorist attacks. Preventive measures are taken on both the domestic and international levels, designed to provide permanent protection from disasters. Not all disasters, particularly natural disasters, can be prevented, but the risk of loss of life and injury can be mitigated with good evacuation plans, environmental planning and design standards. In January 2005, 167 Governments adopted a 10-year global plan for natural disaster risk reduction called the Hyogo Framework.

Preventing or reducing the impacts of disasters on our communities is a key focus for emergency management efforts today. Prevention and mitigation also help reduce the financial costs of disaster response and recovery. Public Safety Canada is working with provincial and territorial governments and stakeholders to promote disaster prevention and mitigation using a risk-based and all-hazards approach. In 2008, Federal/Provincial/Territorial Ministers endorsed a National Disaster Mitigation Strategy.

Mitigation

Preventive or mitigation measures take different forms for different types of disasters. In earthquake prone areas, these preventive measures might include structural changes such as the installation of an earthquake valve to instantly shut off the natural gas supply, seismic retrofits of property, and the securing of items inside a building. The latter may include the mounting of furniture, refrigerators, water heaters and breakables to the walls, and the addition of cabinet latches. In flood prone areas, houses can be built on poles/stilts. In areas prone to prolonged electricity black-outs installation of a generator ensures continuation of electrical service. The construction of storm cellars and fallout shelters are further examples of personal mitigative actions.

On a national level, governments might implement large scale mitigation measures. After the monsoon floods of 2010, the Punjab government subsequently constructed 22 'disaster-resilient' model villages, comprising 1885 single-storey homes, together with schools and health centers.

Disaster mitigation measures are those that eliminate or reduce the impacts and risks of hazards through proactive measures taken before an emergency or disaster occurs.

One of the best known examples of investment in disaster mitigation is the Red River Floodway. The building of the Floodway was a joint provincial/federal undertaking to protect the City of Winnipeg and reduce the impact of flooding in the Red River Basin. It cost $62.7 million to build in the 1960s. Since then, the floodway has been used over 20 times. Its use during the 1997 Red River Flood alone saved an estimated $4.5 billion in costs from potential damage to the city. The Floodway was expanded in 2006 as a joint provincial/federal initiative.

Preparedness

Preparedness focuses on preparing equipment and procedures for use when a disaster occurs.

This equipment and these procedures can be used to reduce vulnerability to disaster, to mitigate the impacts of a disaster or to respond more efficiently in an emergency. The Federal Emergency Management Agency (FEMA) has set out a basic four-stage vision of preparedness flowing from mitigation to preparedness to response to recovery and back to mitigation in a circular planning process. This circular, overlapping model has been modified by other agencies, taught in emergency class and discussed in academic papers. FEMA also operates a Building Science Branch that develops and produces multi-hazard mitigation guidance that focuses on creating disaster-resilient communities to reduce loss of life and property. FEMA advises citizens to prepare their homes with some emergency essentials in the case that the food distribution lines are interrupted. FEMA has subsequently prepared for this contingency by purchasing hundreds of thousands of freeze dried food emergency meals ready to eat (MRE's) to dispense to the communities where emergency shelter and evacuations are implemented.

An airport emergency preparedness exercise

Emergency preparedness can be difficult to measure. CDC focuses on evaluating the effectiveness of its public health efforts through a variety of measurement and assessment programs.

Local Emergency Planning Committees

Local Emergency Planning Committees (LEPCs) are required by the United States Environmental Protection Agency under the Emergency Planning and Community Right-to-Know Act to develop an emergency response plan, review the plan at least annually, and provide information about chemicals in the community to local citizens. This emergency preparedness effort focuses on hazards presented by use and storage of extremely hazardous, hazardous and toxic chemicals. Particular requirements of LEPCs include

- Identification of facilities and transportation routes of extremely hazardous substances
- Description of emergency response procedures, on and off site

- Designation of a community coordinator and facility emergency coordinator(s) to implement the plan
- Outline of emergency notification procedures
- Description of how to determine the probable affected area and population by releases
- Description of local emergency equipment and facilities and the persons responsible for them
- Outline of evacuation plans
- A training program for emergency responders (including schedules)
- Methods and schedules for exercising emergency response plans

According to the EPA, "Many LEPCs have expanded their activities beyond the requirements of EPCRA, encouraging accident prevention and risk reduction, and addressing homeland security in their communities" and the Agency offers advice on how to evaluate the effectiveness of these committees.

Preparedness Measures

Preparedness measures can take many forms ranging from focusing on individual people, locations or incidents to broader, government-based "all hazard" planning. There are a number of preparedness stages between "all hazard' and individual planning, generally involving some combination of both mitigation and response planning. Business continuity planning encourages businesses to have a Disaster Recovery Plan. Community- and faith-based organizations mitigation efforts promote field response teams and inter-agency planning.

Classroom response kit

School-based response teams cover everything from live shooters to gas leaks and nearby bank robberies. Educational institutions plan for cyberattacks and windstorms. Industry specific guidance exists for horse farms, boat owners and more.

Family preparedness for disaster is fairly unusual. A 2013 survey found that only 19% of American families felt that they were "very prepared" for a disaster. Still, there are many resources available for family disaster planning. The Department of Homeland Security's Ready.gov page includes a Family Emergency Plan Checklist, has a whole webpage devoted to readiness for kids, complete with cartoon-style superheroes, and ran a Thunderclap Campaign in 2014. The Center for Disease Control has a Zombie Apocalypse website.

Kitchen fire extinguisher

Disasters take a variety of forms to include earthquakes, tsunamis or regular structure fires. That a disaster or emergency is not large scale in terms of population or acreage impacted or duration does not make it any less of a disaster for the people or area impacted and much can be learned about preparedness from so-called small disasters. The Red Cross states that it responds to nearly 70,000 disasters a year, the most common of which is a single family fire.

Items on shelves in basement

Preparedness starts with an individual's everyday life and involves items and training that would be useful in an emergency. What is useful in an emergency is often also useful in everyday life. From personal preparedness, preparedness continues on a continuum through family preparedness, community preparedness and then business, non-profit and governmental preparedness. Some organizations blend these various levels. For example, the International Red Cross and Red Crescent Movement has a webpage on disaster training as well as offering training on basic preparedness such as Cardiopulmonary resuscitation and First Aid. Other non-profits such as Team Rubicon bring specific groups of people into disaster preparedness and response operations. FEMA breaks down preparedness into a pyramid, with citizens on the foundational bottom, on top of which rests local government, state government and federal government in that order.

Non-perishable food in cabinet

The basic theme behind preparedness is to be ready for an emergency and there are a number of different variations of being ready based on an assessment of what sort of threats exist. Nonetheless, there is basic guidance for preparedness that is common despite an area's specific dangers. FEMA recommends that everyone have a three-day survival kit for their household. Because individual household sizes and specific needs might vary, FEMA's recommendations are not item specific, but the list includes:

- Three-day supply of non-perishable food.
- Three-day supply of water – one gallon of water per person, per day.
- Portable, battery-powered radio or television and extra batteries.
- Flashlight and extra batteries.

- First aid kit and manual.
- Sanitation and hygiene items (e.g. toilet paper, menstrual hygiene products).
- Matches and waterproof container.
- Whistle.
- Extra clothing.
- Kitchen accessories and cooking utensils, including a can opener.
- Photocopies of credit and identification cards.
- Cash and coins.
- Special needs items, such as prescription medications, eyeglasses, contact lens solutions, and hearing aid batteries.
- Items for infants, such as formula, diapers, bottles, and pacifiers.
- Other items to meet unique family needs.

Along similar lines, but not exactly the same, CDC has its own list for a proper disaster supply kit.

- Water—one gallon per person, per day
- Food—nonperishable, easy-to-prepare items
- Flashlight
- Battery powered or hand crank radio (NOAA Weather Radio, if possible)
- Extra batteries
- First aid kit
- Medications (7-day supply), other medical supplies, and medical paperwork (e.g., medication list and pertinent medical information)
- Multipurpose tool (e.g., Swiss army knife)
- Sanitation and personal hygiene items
- Copies of personal documents (e.g., proof of address, deed/lease to home, passports, birth certificates, and insurance policies)
- Cell phone with chargers
- Family and emergency contact information
- Extra cash
- Emergency blanket
- Map(s) of the area
- Extra set of car keys and house keys
- Manual can opener

Children are a special population when considering Emergency preparedness and many resources are directly focused on supporting them. SAMHSA has list of tips for talking to children during infectious disease outbreaks, to include being a good listener, encouraging children to ask questions and modeling self-care by setting routines, eating healthy meals, getting enough sleep and taking deep breaths to handle stress. FEMA has similar advice, noting that "Disasters can leave children feeling frightened, confused, and insecure" whether a child has experienced it first hand, had it happen to a friend or simply saw it on television. In the same publication, FEMA further notes, "Preparing for disaster helps everyone in the family accept the fact that disasters do happen, and provides an opportunity to identify and collect the resources needed to meet basic needs after disaster. Preparation helps; when people feel prepared, they cope better and so do children."

To help people assess what threats might be in order to augment their emergency supplies or improve their disaster response skills, FEMA has published a booklet called the "Threat and Hazard Identification and Risk Assessment Guide." (THIRA) This guide, which outlines the THIRA process, emphasizes "whole community involvement," not just governmental agencies, in preparedness efforts. In this guide, FEMA breaks down hazards into three categories: Natural, technological and human caused and notes that each hazard should be assessed for both its likelihood and its significance. According to FEMA, "Communities should consider only those threats and hazards that could plausibly occur" and "Communities should consider only those threats and hazards that would have a significant effect on them." To develop threat and hazard context descriptions, communities should take into account the time, place, and conditions in which threats or hazards might occur.

Not all preparedness efforts and discussions involve the government or established NGOs like the Red Cross. Emergency preparation discussions are active on the internet, with many blogs and websites dedicated to discussing various aspects of preparedness. On-line sales of items such as survival food, medical supplies and heirloom seeds allow people to stock basements with cases of food and drinks with 25 year shelf lives, sophisticated medical kits and seeds that are guaranteed to sprout even after years of storage.

One group of people who put a lot of effort in disaster preparations is called Doomsday Preppers. This subset of preparedness-minded people often share a belief that the FEMA or Red Cross emergency preparation suggestions and training are not extensive enough. Sometimes called survivalists, Doomsday Preppers are often preparing for The End Of The World As We Know It, abbreviated as TEOTWAWKI. With a motto some have that "The Future Belongs to those who Prepare," this Preparedness subset has its own set of Murphy's Rules, including "Rule Number 1: Food, you still don't have enough" and "Rule Number 26: People who thought the Government would save them, found out that it didn't."

Not all emergency preparation efforts revolve around food, guns and shelters, though these items help address the needs in the bottom two sections of Maslow's hierarchy of needs. The American Preppers Network has an extensive list of items that might be useful in less apparent ways than a first aid kid or help add 'fun' to challenging times. These items include:

- Books and magazines
- Arts and crafts painting
- Children's entertainment
- Crayons and coloring books

- Notebooks and writing supplies
- Nuts, bolts, screws, nails, etc.
- Religious material
- Sporting equipment, card games and board games
- Posters and banners creating awareness

Emergency preparedness goes beyond immediate family members. For many people, pets are an integral part of their families and emergency preparation advice includes them as well. It is not unknown for pet owners to die while trying to rescue their pets from a fire or from drowning. CDC's Disaster Supply Checklist for Pets includes:

- Food and water for at least 3 days for each pet; bowls, and a manual can opener.
- Depending on the pet you may need a litter box, paper towels, plastic trash bags, grooming items, and/or household bleach.
- Medications and medical records stored in a waterproof container.
- First aid kit with a pet first aid book.
- Sturdy leash, harness, and carrier to transport pet safely. A carrier should be large enough for the animal to stand comfortably, turn around, and lie down. Your pet may have to stay in the carrier for several hours.
- Pet toys and the pet's bed, if you can easily take it, to reduce stress.
- Current photos and descriptions of your pets to help others identify them in case you and your pets become separated, and to prove that they are yours.
- Information on feeding schedules, medical conditions, behavior problems, and the name and telephone number of your veterinarian in case you have to board your pets or place them in foster care.

Emergency preparedness also includes more than physical items and skill-specific training. Psychological preparedness is also a type of emergency preparedness and specific mental health preparedness resources are offered for mental health professionals by organizations such as the Red Cross. These mental health preparedness resources are designed to support both community members affected by a disaster and the disaster workers serving them. CDC has a website devoted to coping with a disaster or traumatic event. After such an event, the CDC, through the Substance Abuse and Mental Health Services Administration (SAMHSA), suggests that people seek psychological help when they exhibit symptoms such as excessive worry, crying frequently, an increase in irritability, anger, and frequent arguing, wanting to be alone most of the time, feeling anxious or fearful, overwhelmed by sadness, confused, having trouble thinking clearly and concentrating, and difficulty making decisions, increased alcohol and/or substance use, increased physical (aches, pains) complaints such as headaches and trouble with "nerves."

Sometimes emergency supplies are kept in what is called a Bug-out bag. While FEMA does not actually use the term "Bug out bag," calling it instead some variation of a "Go Kit," the idea of having emergency items in a quickly accessible place is common to both FEMA and CDC, though on-line discussions of what items a "bug out bag" should include sometimes cover items such as firearms and great knives that are not specifically suggested by FEMA or CDC. The theory behind a "bug out bag" is that emergency preparations should include the possibility of Emergency evacuation. Whether fleeing a burning building or hastily packing a car to escape an impending hurricane, flood or dangerous chemical release, rapid departure from a home or workplace environment is always a possibility and FEMA suggests having a Family Emergency Plan for such occasions. Because family members may not be together when disaster strikes, this plan should include reliable contact information for friends or relatives who live outside of what would be the disaster area for household members to notify they are safe or otherwise communicate with each other. Along with the contact information, FEMA suggests having well-understood local gathering points if a house must be evacuated quickly to avoid the dangers of re-reentering a burning home. Family and emergency contact information should be printed on cards and put in each family member's backpack or wallet. If family members spend a significant amount of time in a specific location, such as at work or school, FEMA suggests learning the emergency preparation plans for those places. FEMA has a specific form, in English and in Spanish, to help people put together these emergency plans, though it lacks lines for email contact information.

Like children, people with disabilities and other special needs have special emergency preparation needs. While "disability" has a specific meaning for specific organizations such as collecting Social Security benefits, for the purposes of emergency preparedness, the Red Cross uses the term in a broader sense to include people with physical, medical, sensor or cognitive disabilities or the elderly and other special needs populations. Depending on the particular disability, specific emergency preparations might be required. FEMA's suggestions for people with disabilities includes having copies of prescriptions, charging devices for medical devices such as motorized wheel chairs and a week's supply of medication readily available LINK or in a "go stay kit." In some instances, lack of competency in English may lead to special preparation requirements and communication efforts for both individuals and responders.

FEMA notes that long term power outages can cause damage beyond the original disaster that can be mitigated with emergency generators or other power sources to provide an Emergency power system. The United States Department of Energy states that 'homeowners, business owners, and local leaders may have to take an active role in dealing with energy disruptions on their own." This active role may include installing or other procuring generators that are either portable or permanently mounted and run on fuels such as propane or natural gas or gasoline. Concerns about carbon monoxide poisoning, electrocution, flooding, fuel storage and fire lead even small property owners to consider professional installation and maintenance. Major institutions like hospitals, military bases and educational institutions often have or are considering extensive backup power systems. Instead of, or in addition to, fuel-based power systems, solar, wind and other alternative power sources may be used. Standalone batteries, large or small, are also used to provide backup charging for electrical systems and devices ranging from emergency lights to computers to cell phones.

Emergency preparedness does not stop at home or at school. The United States Department of Health and Human Services addresses specific emergency preparedness issues hospitals may have to respond to, including maintaining a safe temperature, providing adequate electricity for life support systems and even carrying out evacuations under extreme circumstances. FEMA encourages all businesses to have businesses to have an emergency response plan and the Small Business Administration specifically advises small business owners to also focus emergency preparedness and provides a variety of different worksheets and resources.

FEMA cautions that emergencies happen while people are travelling as well and provides guidance around emergency preparedness for a range travelers to include commuters, *Commuter Emergency Plan* and holiday travelers. In particular, Ready.gov has a number of emergency preparations specifically designed for people with cars. These preparations include having a full gas tank, maintaining adequate windshield wiper fluid and other basic car maintenance tips. Items specific to an emergency include:

- Jumper cables: might want to include flares or reflective triangle
- Flashlights, to include extra batteries (batteries have less power in colder weather)
- First Aid Kit, to include any necessary medications, baby formula and diapers if caring for small children
- Non-perishable food such as canned food (be alert to liquids freezing in colder weather), and protein rich foods like nuts and energy bars
- Manual can opener
- At least 1 gallon of water per person a day for at least 3 days (be alert to hazards of frozen water and resultant container rupture)
- Basic toolkit: pliers, wrench, screwdriver
- Pet supplies: food and water
- Radio: battery or hand cranked
- For snowy areas: cat litter or sand for better tire traction; shovel; ice scraper; warm clothes, gloves, hat, sturdy boots, jacket and an extra change of clothes
- Blankets or sleeping bags
- Charged Cell Phone: and car charger

In addition to emergency supplies and training for various situations, FEMA offers advice on how to mitigate disasters. The Agency gives instructions on how to retrofit a home to minimize hazards from a Flood, to include installing a Backflow prevention device, anchoring fuel tanks and relocating electrical panels.

Given the explosive danger posed by natural gas leaks, Ready.gov states unequivocally that "It is vital that all household members know how to shut off natural gas" and that property owners must ensure they have any special tools needed for their particular gas hookups. Ready.gov also notes that "It is wise to teach all responsible household members where and how to shut off the electricity," cautioning that individual circuits should be shut off before the main circuit. Ready.gov further states that "It is vital that all household members learn how to shut off the water at the main house valve" and cautions that the possibility that rusty valves might require replacement.

Marked gas shuttoff

Response

The response phase of an emergency may commence with Search and Rescue but in all cases the focus will quickly turn to fulfilling the basic humanitarian needs of the affected population. This assistance may be provided by national or international agencies and organizations. Effective coordination of disaster assistance is often crucial, particularly when many organizations respond and local emergency management agency (LEMA) capacity has been exceeded by the demand or diminished by the disaster itself. The National Response Framework is a United States government publication that explains responsibilities and expectations of government officials at the local, state, federal, and tribal levels. It provides guidance on Emergency Support Functions that may be integrated in whole or parts to aid in the response and recovery process.

On a personal level the response can take the shape either of a *shelter in place* or an *evacuation*.

In a shelter-in-place scenario, a family would be prepared to fend for themselves in their home for many days without any form of outside support. In an *evacuation*, a family leaves the area by automobile or other mode of transportation, taking with them the maximum amount of supplies they can carry, possibly including a tent for shelter. If mechanical transportation is not available, evacuation on foot would ideally include carrying at least three days of supplies and rain-tight bedding, a tarpaulin and a bedroll of blankets.

Donations are often sought during this period, especially for large disasters that overwhelm local capacity. Due to efficiencies of scale, money is often the most cost-effective donation if fraud is avoided. Money is also the most flexible, and if goods are sourced locally then transportation is minimized and the local economy is boosted. Some donors prefer to send gifts in kind, however these items can end up creating issues, rather than helping. One innovation by Occupy Sandy volunteers is to use a donation registry, where families and businesses impacted by the disaster can make specific requests, which remote donors can purchase directly via a web site.

Evacuation sign

Medical considerations will vary greatly based on the type of disaster and secondary effects. Survivors may sustain a multitude of injuries to include lacerations, burns, near drowning, or crush syndrome.

Recovery

The recovery phase starts after the immediate threat to human life has subsided. The immediate goal of the recovery phase is to bring the affected area back to normalcy as quickly as possible. During reconstruction it is recommended to consider the location or construction material of the property.

The most extreme home confinement scenarios include war, famine and severe epidemics and may last a year or more. Then recovery will take place inside the home. Planners for these events usually buy bulk foods and appropriate storage and preparation equipment, and eat the food as part of normal life. A simple balanced diet can be constructed from vitamin pills, whole-meal wheat, beans, dried milk, corn, and cooking oil. One should add vegetables, fruits, spices and meats, both prepared and fresh-gardened, when possible.

As a Profession

Professional emergency managers can focus on government and community preparedness, or private business preparedness. Training is provided by local, state, federal and private organizations and ranges from public information and media relations to high-level incident command and tactical skills.

In the past, the field of emergency management has been populated mostly by people with a military or first responder background. Currently, the field has become more diverse, with many managers coming from a variety of backgrounds other than the military or first responder fields. Educational opportunities are increasing for those seeking undergraduate and graduate degrees in emergency management or a related field. There are over 180 schools in the US with emergency management-related programs, but only one doctoral program specifically in emergency management.

Professional certifications such as Certified Emergency Manager (CEM) and Certified Business Continuity Professional (CBCP) are becoming more common as professional standards are raised throughout the field, particularly in the United States. There are also professional organizations for emergency managers, such as the National Emergency Management Association and the International Association of Emergency Managers.

Principles

In 2007, Dr. Wayne Blanchard of FEMA's Emergency Management Higher Education Project, at the direction of Dr. Cortez Lawrence, Superintendent of FEMA's Emergency Management Institute, convened a working group of emergency management practitioners and academics to consider principles of emergency management. This was the first time the principles of the discipline were to be codified. The group agreed on eight principles that will be used to guide the development of a doctrine of emergency management. Below is a summary:

- Comprehensive – consider and take into account all hazards, all phases, all stakeholders and all impacts relevant to disasters.
- Progressive – anticipate future disasters and take preventive and preparatory measures to build disaster-resistant and disaster-resilient communities.
- Risk-driven – use sound risk management principles (hazard identification, risk analysis, and impact analysis) in assigning priorities and resources.
- Integrated – ensure unity of effort among all levels of government and all elements of a community.
- Collaborative – create and sustain broad and sincere relationships among individuals and organizations to encourage trust, advocate a team atmosphere, build consensus, and facilitate communication.
- Coordinated – synchronize the activities of all relevant stakeholders to achieve a common purpose.
- Flexible – use creative and innovative approaches in solving disaster challenges.
- Professional – value a science and knowledge-based approach; based on

education, training, experience, ethical practice, public stewardship and continuous improvement.

A fuller description of these principles can be found at

Tools

In recent years the continuity feature of emergency management has resulted in a new concept, Emergency Management Information Systems (EMIS). For continuity and interoperability between emergency management stakeholders, EMIS supports an infrastructure that integrates emergency plans at all levels of government and non-government involvement for all four phases of emergencies. In the healthcare field, hospitals utilize the Hospital Incident Command System (HICS), which provides structure and organization in a clearly defined chain of command.

Disaster Response Technologies

Smart Emergency Response System (SERS) prototype was built in the SmartAmerica Challenge 2013-2014, a United States government initiative. SERS has been created by a team of nine organizations led by MathWorks. The project was featured at the White House in June 2014 and described by Todd Park (U.S. Chief Technology Officer) as an exemplary achievement.

The Smart America initiative challenges the participants to build cyber-physical systems as a glimpse of the future to save lives, create jobs, foster businesses, and improve the economy. SERS primarily saves lives. The system provides the survivors and the emergency personnel with information to locate and assist each other during a disaster. SERS allows to submit help requests to a MATLAB-based mission center connecting first responders, apps, search-and-rescue dogs, a 6-feet-tall humanoid, robots, drones, and autonomous aircraft and ground vehicles. The command and control center optimizes the available resources to serve every incoming requests and generates an action plan for the mission. The Wi-Fi network is created on the fly by the drones equipped with antennas. In addition, the autonomous rotorcrafts, planes, and ground vehicles are simulated with Simulink and visualized in a 3D environment (Google Earth) to unlock the ability to observe the operations on a mass scale.

Within other Professions

Practitioners in emergency management come from an increasing variety of backgrounds. Professionals from memory institutions (e.g., museums, historical societies, etc.) are dedicated to preserving cultural heritage—objects and records. This has been an increasingly major component within this field as a result of the heightened awareness following the September 11 attacks in 2001, the hurricanes in 2005, and the collapse of the Cologne Archives.

A disaster plan book at Rockefeller University in a biochemistry research laboratory.

To increase the potential successful recovery of valuable records, a well-established and thoroughly tested plan must be developed. This plan should emphasize simplicity in order to aid in response and recovery: employees should perform similar tasks in the response and recovery phase that they perform under normal conditions. It should also include mitigation strategies such as the installation of sprinklers within the institution. Professional associations hold regular workshops to keep individuals up to date with tools and resources in order to minimize risk and maximize recovery.

Other Tools

In 2008, the U.S. Agency for International Development created a web-based tool for estimating populations impacted by disasters. Called Population Explorer the tool uses land scan population data, developed by Oak Ridge National Laboratory, to distribute population at a resolution 1 km^2 for all countries in the world. Used by USAID's FEWS NET Project to estimate populations vulnerable and or imd by food insecurity, Population Explorer is gaining wide use in a range of emergency analysis and response actions, including estimating populations impacted by floods in Central America and the Pacific Ocean Tsunami event in 2009.

In 2007, a checklist for veterinarians was published in the Journal of the American Veterinary Medical Association, it had two sets of questions for a professional to ask themselves before assisting with an emergency:

Absolute requirements for participation:

- Have I chosen to participate?
- Have I taken ICS training?
- Have I taken other required background courses?
- Have I made arrangements with my practice to deploy?
- Have I made arrangements with my family?

Incident Participation:

- Have I been invited to participate
- Are my skill sets a match for the mission?
- Can I access just-in-time training to refresh skills or acquire needed new skills?
- Is this a self-support mission?
- Do I have supplies needed for three to five days of self-support?

While written for veterinarians, this checklist is applicable for any professional to consider before assisting with an emergency.

INTERNATIONAL ORGANIZATIONS

The International Emergency Management Society

The International Emergency Management Society (TIEMS), is an international non-profit NGO, registered in Belgium. TIEMS is a Global Forum for Education, Training, Certification and Policy in Emergency and Disaster Management. TIEMS' goal is to develop and bring modern emergency management tools, and techniques into practice, through the exchange of information, methodology innovations and new technologies.

TIEMS provides a platform for stakeholders to meet, network and learn about new technical and operational methodologies. TIEMS focuses on cultural differences to be understood and included in the society's events, education and research programs. This is achieved by establishing local chapters worldwide. Today, TIEMS has chapters in Benelux, Romania, Finland, Italy, Middle East and North Africa (MENA), Iraq, India, Korea, Japan and China.

International Association of Emergency Managers

The International Association of Emergency Managers (IAEM) is a non-profit educational organization aimed at promoting the goals of saving lives and property protection during emergencies. The mission of IAEM is to serve its members by providing information, networking and professional opportunities, and to advance the emergency management profession.

It has seven councils around the world: Asia, Canada, Europa, International, Oceania, Student and USA.

The Air Force Emergency Management Association, affiliated by membership with the IAEM, provides emergency management information and networking for U.S. Air Force Emergency Management personnel.

International Recovery Platform

The International Recovery Platform (IRP) was conceived at the World Conference on Disaster Reduction (WCDR) in Kobe, Hyogo, Japan in January 2005, as part of the

Hyogo Framework for Action (HFA) 2005–2015. The HFA is a global plan for disaster risk reduction adopted by 168 governments.

The key role of IRP is to identify gaps in post disaster recovery and to serve as a catalyst for the development of tools and resources for recovery efforts.

The International Red Cross and Red Crescent Movement

The International Federation of Red Cross and Red Crescent Societies (IFRC) works closely with National Red Cross and Red Crescent societies in responding to emergencies, many times playing a pivotal role. In addition, the IFRC may deploy assessment teams, e.g. Field Assessment and Coordination Teams (FACT), to the affected country if requested by the national society. After assessing the needs, Emergency Response Units (ERUs) may be deployed to the affected country or region. They are specialized in the response component of the emergency management framework.

Baptist Global Response

Baptist Global Response (BGR) is a disaster relief and community development organization. BGR and its partners respond globally to people with critical needs worldwide, whether those needs arise from chronic conditions or acute crises such as natural disasters. While BGR is not an official entity of the Southern Baptist Convention, it is rooted in Southern Baptist life and is the international partnership of Southern Baptist Disaster Relief teams, which operate primarily in the US and Canada.

United Nations

The United Nations system rests with the Resident Coordinator within the affected country. However, in practice, the UN response will be coordinated by the UN Office for the Coordination of Humanitarian Affairs (UN-OCHA), by deploying a UN Disaster Assessment and Coordination (UNDAC) team, in response to a request by the affected country's government.

World Bank

Since 1980, the World Bank has approved more than 500 projects related to disaster management, dealing with both disaster mitigation as well as reconstruction projects, amounting to more than US$40 billion. These projects have taken place all over the world, in countries such as Argentina, Bangladesh, Colombia, Haiti, India, Mexico, Turkey and Vietnam.

Prevention and mitigation projects include forest fire prevention measures, such as early warning measures and education campaigns; early-warning systems for hurricanes; flood prevention mechanisms (e.g. shore protection, terracing, etc.); and earthquake-prone construction. In a joint venture with Columbia University under the umbrella of the ProVention Consortium the World Bank has established a Global Risk Analysis of Natural

Disaster Hotspots. In June 2006, the World Bank, in response to the HFA, established the Global Facility for Disaster Reduction and Recovery (GFDRR), a partnership with other aid donors to reduce disaster losses. GFDRR helps developing countries fund development projects and programs that enhance local capacities for disaster prevention and emergency preparedness.

European Union

In 2001 the EU adopted Community Mechanism for Civil Protection, to facilitate co-operation in the event of major emergencies requiring urgent response actions. This also applies to situations where there may be an imminent threat as well.

The heart of the Mechanism is the Monitoring and Information Center (MIC), part of the European Commission's Directorate-General for Humanitarian Aid & Civil Protection. Accessible 24 hours a day, it gives countries access to a one-stop-shop of civil protections available amongst all the participating states. Any country inside or outside the Union affected by a major disaster can make an appeal for assistance through the MIC. It acts as a communication hub, and provides useful and updated information on the actual status of an ongoing emergency.

NATIONAL ORGANIZATIONS

Australia

Natural disasters are part of life in Australia. Heatwaves have killed more Australians than any other type of natural disaster in the 20th century. Australia's emergency management processes embrace the concept of the prepared community. The principal government agency in achieving this is Emergency Management Australia.

Canada

Public Safety Canada is Canada's national emergency management agency. Each province is required to have both legislation for dealing with emergencies, and provincial emergency management agencies, typically called "Emergency Measures Organizations" (EMO). Public Safety Canada co-ordinates and supports the efforts of federal organizations as well as other levels of government, first responders, community groups, the private sector, and other nations. The Public Safety and Emergency Preparedness Act defines the powers, duties and functions of PS are outlined. Other acts are specific to individual fields such as corrections, law enforcement, and national security.

Germany

In Germany the Federal Government controls the German *Katastrophenschutz* (disaster relief), the Technisches Hilfswerk (*Federal Agency for Technical Relief*, THW), and the *Zivilschutz* (civil protection) programs coordinated by the *Federal Office of Civil*

Protection and Disaster Assistance. Local fire department units, the German Armed Forces (Bundeswehr), the German Federal Police and the 16 state police forces (Länderpolizei) are also deployed during disaster relief operations.

There are several private organizations in Germany that also deal with emergency relief. Among these are the German Red Cross, Johanniter-Unfall-Hilfe (the German equivalent of the St. John Ambulance), the Malteser-Hilfsdienst, and the Arbeiter-Samariter-Bund. As of 2006, there is a program of study at the University of Bonn leading to the degree "Master in Disaster Prevention and Risk Governance" As a support function radio amateurs provide additional emergency communication networks with frequent trainings.

India

The National Disaster Management Authority is the primary government agency responsible for planning and capacity-building for disaster relief. Its emphasis is primarily on strategic risk management and mitigation, as well as developing policies and planning. The National Institute of Disaster Management is a policy think-tank and training institution for developing guidelines and training programs for mitigating disasters and managing crisis response.

A protective wall built on the shore of the coastal town of Kalpakkam, in aftermath of the 2004 Indian Ocean earthquake.

The National Disaster Response Force is the government agency primarily responsible for emergency management during natural and man-made disasters, with specialized skills in search, rescue and rehabilitation. The Ministry of Science and Technology also contains an agency that brings the expertise of earth scientists and meteorologists to emergency management. The Indian Armed Forces also plays an important role in the rescue/recovery operations after disasters. Aniruddha's Academy of Disaster Management (ACDM) is a non-profit organization in Mumbai, India with 'disaster management' as its principal objective.

Malaysia

In Malaysia, The National Security Council has the responsibility to handle emergency and disaster events. Ministry of Home Affairs Malaysia, Ministry of Health Malaysia and Ministry of Housing, Urban Wellbeing and Local Government Malaysia are also having responsibility in managing emergency. Several agencies are involved in emergency managements are Royal Malaysian Police, Malaysian Fire and Rescue Department, Malaysian Civil Defence Force, Ministry of Health Malaysia and Malaysian Maritime Enforcement Agency. There were also some voluntary organisation who involved themselves in emergency/ disaster management such as St. John Ambulance of Malaysia, Malaysian Red Crescent Society and so on.

New Zealand

In New Zealand, depending on the scope of the emergency/disaster, responsibility may be handled at either the local or national level. Within each region, local governments are organized into 16 Civil Defence Emergency Management Groups (CMGs). If local arrangements are overwhelmed, pre-existing mutual-support arrangements are activated. Central government has the authority to coordinate the response through the National Crisis Management Centre (NCMC), operated by the Ministry of Civil Defence & Emergency Management (MCDEM). These structures are defined by regulation, and explained in *The Guide to the National Civil Defence Emergency Management Plan 2006*, roughly equivalent to the U.S. Federal Emergency Management Agency's National Response Framework.

New Zealand uses unique terminology for emergency management. Emergency management is rarely used, many government publications retaining the use of the term civil defence. For example, the Minister of Civil Defence is responsible for the MCDEM. Civil Defence Emergency Management is a term in its own right, defined by statute. And disaster rarely appears in official publications, emergency and incident being the preferred terms, with the term event also being used. For example, publications refer to the Canterbury Snow Event 2002.

"4Rs" is the emergency management cycle used in New Zealand, its four phases are known as:

- Reduction = Mitigation
- Readiness = Preparedness
- Response
- Recovery

Pakistan

Disaster management in Pakistan revolves around flood disasters focusing on rescue and relief. There is a dearth of knowledge and information about hazard identification, risk assessment and management, and disaster preparedness. Disaster management, development planning and environmental management institutions operate in isolation with no integrated planning, there being no central authority for integrated disaster management. State-level measures are heavily tilted towards structural aspects.

Russia

In Russia, the Ministry of Emergency Situations (EMERCOM) is engaged in fire fighting, civil defense, and search and rescue after both natural and human-made disasters.

Somalia

In Somalia, the Federal Government announced in May 2013 that the Cabinet had approved draft legislation on a new Somali Disaster Management Agency (SDMA), which had originally been proposed by the Ministry of Interior. According to the Prime Minister's Media Office, the SDMA will lead and coordinate the government's response to various natural disasters. It is part of a broader effort by the federal authorities to re-establish national institutions. The Federal Parliament is now expected to deliberate on the proposed bill for endorsement after any amendments.

The Netherlands

In the Netherlands the Ministry of Security and Justice is responsible for emergency preparedness and emergency management on a national level and operates a national crisis Centre (NCC). The country is divided into 25 safety regions (veiligheidsregio). In a safety region, there are four components: the regional fire department, the regional department for medical care(ambulances and psycho-sociological care etc.), the regional dispatch and a section for risk- and crisis management. The regional dispatch operates for police, fire department and the regional medical care. The dispatch has all these three services combined into one dispatch for the best multi-coordinated response to an incident or an emergency. And also facilitates in information management, emergency communication and care of citizens. These services are the main structure for a response to an emergency. It can happen that, for a specific emergency, the co-operation with an other service is needed, for instance the Ministry of Defence, water board(s) or Rijkswaterstaat. The veiligheidsregio can integrate these other services into their structure by adding them to specific conferences on operational or administrative level.

All regions operate according to the Coordinated Regional Incident Management system.

United Kingdom

Following the 2000 fuel protests and severe flooding that same year, as well as the foot-and-mouth crisis in 2001, the United Kingdom passed the Civil Contingencies Act 2004 (CCA). The CCA defined some organizations as Category 1 and 2 Responders, setting responsibilities regarding emergency preparedness and response. It is managed by the Civil Contingencies Secretariat through Regional Resilience Forums and local authorities.

Disaster Management training is generally conducted at the local level, and consolidated through professional courses that can be taken at the Emergency Planning College. Diplomas, undergraduate and postgraduate qualifications can be gained at universities throughout the country. The Institute of Emergency Management is a charity, established in 1996, providing consulting services for the government, media and commercial sectors. There are a number of professional societies for Emergency Planners including the Emergency Planning Society and the Institute of Civil Protection and Emergency Management.

One of the largest emergency exercises in the UK was carried out on 20 May 2007 near Belfast, Northern Ireland: a simulated plane crash-landing at Belfast International Airport. Staff from five hospitals and three airports participated in the drill, and almost 150 international observers assessed its effectiveness.

United States

Disaster management in the United States has utilized the functional All-Hazards approach for over 20 years, in which managers develop processes (such as communication & warning or sheltering) rather than developing single-hazard or threat focused plans (e.g., a tornado plan). Processes are then mapped to specific hazards or threats, with the manager looking for gaps, overlaps, and conflicts between processes.

Given these notions, emergency managers must identify, contemplate, and assess possible man-made threats and natural threats that may affect their respective locales. Because of geographical differences throughout the nation, a variety of different threats accept communities among the states. Thus, although similarities may exist, no two emergency plans will be completely identical. Additionally, each locale has different resources and capacities (e.g., budgets, personnel, equipment, etc.) for dealing with emergencies. Each individual community must craft its own unique emergency plan that addresses potential threats that are specific to the locality.

This creates a plan more resilient to unique events because all common processes are defined, and encourages planning done by the stakeholders who are closer to the individual processes, such as a traffic management plan written by public works director. This type of planning can lead to conflict with non-emergency management regulatory bodies, which require development of hazard/threat specific plans, such as development of specific H1N1 flu plans and terrorism-specific plans.

In the United States, all disasters are initially local, with local authorities, with usually a police, fire, or EMS agency, taking charge. Many local municipalities may also have a separate dedicated office of emergency management (OEM), along with personnel and equipment. If the event becomes overwhelming to local government, state emergency management (the primary government structure of the United States) becomes the controlling emergency management agency. Federal Emergency Management Agency (FEMA), part of the Department of Homeland Security (DHS), is lead federal agency for emergency management. The United States and its territories are broken down into ten regions for FEMA's emergency management purposes. FEMA supports, but does not override, state authority. The Citizen Corps is an organization of volunteer service programs, administered locally and coordinated nationally by DHS, which seek to mitigate disasters and prepare the population for emergency response through public education, training, and outreach.

Most disaster response is carried out by volunteer organizations. In the US, the Red Cross is chartered by Congress to coordinate disaster response services, including typically being the lead agency handling shelter and feeding of evacuees. Religious organizations, with their ability to provide volunteers quickly, are usually integral during the response process. The largest being the Salvation Army, with a primary focus on chaplaincy and rebuilding, and Southern Baptists who focus on food preparation and distribution, as well as cleaning up after floods and fires, chaplaincy, mobile shower units, chainsaw crews and more. With over 65,000 trained volunteers Southern Baptist Disaster Relief is one of the largest disaster relief organizations in the US. Similar services are also provided by Methodist Relief Services, the Lutherans, and Samaritan's Purse. Unaffiliated volunteers show up at most large disasters. To prevent abuse by criminals and for the safety of the volunteers, procedures have been implemented within most response agencies to manage and effectively use these 'SUVs' (Spontaneous Unaffiliated Volunteers).

The US Congress established the Center for Excellence in Disaster Management and Humanitarian Assistance (COE) as the principal agency to promote disaster preparedness in the Asia-Pacific region.

The National Tribal Emergency Management Council (NEMC) is a non-profit educational organization developed for Tribal organizations to share information and best practices, as well as discussing issues regarding public health and safety, emergency management and homeland security, affecting those under Indian sovereignty. NTMC is organized into Regions, based on the FEMA 10 region system. NTMC was founded by the Northwest Tribal Emergency Management Council (NWTEMC), a consortium of 29 Tribal Nations and Villages in Washington, Idaho, Oregon and Alaska.

If a disaster or emergency is declared to be terror related or an "Incident of National Significance", the Secretary of Homeland Security will initiate the National Response Framework (NRF). The NRF allows the integration of federal resources with local, country, state, or tribal entities, with management of those resources to be handled at the lowest possible level, utilizing the National Incident Management System (NIMS).

FEMA's Emergency Management Institute

The Emergency Management Institute (EMI) serves as the national focal point for the development and delivery of emergency management training to enhance the capabilities of state, territorial, local, and tribal government officials; volunteer organizations; FEMA's disaster workforce; other Federal agencies; and the public and private sectors to minimize the impact of disasters and emergencies on the American public. EMI curricula are structured to meet the needs of this diverse audience with an emphasis on separate organizations working together in all-hazards emergencies to save lives and protect property. Particular emphasis is placed on governing doctrine such as the National Response Framework (NRF), National Incident Management System (NIMS), and the National Preparedness Guidelines. EMI is fully accredited by the International Association for Continuing Education and Training (IACET) and the American Council on Education (ACE).

Emergency Management Institute's Main Campus in Emmitsburg, Maryland

Approximately 5,500 participants attend resident courses each year while 100,000 individuals participate in non-resident programs sponsored by EMI and conducted by state emergency management agencies under cooperative agreements with FEMA. Another 150,000 individuals participate in EMI-supported exercises, and approximately 1,000 individuals participate in the Chemical Stockpile Emergency Preparedness Program (CSEPP).

The *independent study* program at EMI consists of free courses offered to United States citizens in Comprehensive Emergency Management techniques. Course IS-1 is entitled "Emergency Manager: An Orientation to the Position" and provides background information on FEMA and the role of emergency managers in agency and volunteer organization coordination. The EMI Independent Study (IS) Program, a Web-based distance learning program open to the public, delivered extensive online training with approximately 200 courses and trained more than 2.8 million individuals. The EMI IS Web site receives 2.5 to 3 million visitors a day.

REFERENCES

1. "Maine Emergency Management Agency" (2007). "What is Emergency Management?". Retrieved 2014-02-22.
2. Drabek, Thomas (1991). Emergency Management: Principles and Practice for Local Government. Washington, D.C.: International City Management Association. pp. xvii.
3. NEWS: Pakistan's Punjab builds model villages to withstand disasters, Climate & Development Knowledge Network, 17 December 2013.
4. Simonovic, Slobodan. "Flood Mitigation Efforts in the Red River Basin" (PDF). www.iclr.org. The Institute of Catastrophic Loss Reduction. Retrieved 20 October 2016.
5. "Animals in Disasters". Training.fema.gov. Retrieved 2015-03-06.

6. Baird, Malcolm E. (2010). "The "Phases" of Emergency Management" (PDF). Vanderbilt Center for Transportation Research. Retrieved 2015-03-08.
7. "Building Science". fema.gov. Retrieved 8 March 2015.
8. "Rand Homeland Security" (PDF). Rand.org. Retrieved 2015-03-08.
9. "Public Health Emergency Preparedness Cooperative Agreement" (PDF). Cdc.gov. Retrieved 2015-03-08.
10. "US Environmental Protection Agency | US EPA". .epa.gov. 2015-01-28. Retrieved 2015-03-08.
11. "The Emergency Planning and Community Right-to-Know Act" (PDF). Epa.gov. Retrieved 2015-03-08.
12. "Masuring Progress in Chemical Safety : A Guide for Local Emergency Planning Committees and Similar Groups" (PDF). Epa.gov. Retrieved 2015-03-08.
13. http://www.fema.gov/pdf/plan/slg101.pdf Guide for All-Hazard Emergency Operations Planning September 1996
14. "Community-Based Pre-Disaster Mitigation" (PDF). Fema.gov. Retrieved 2015-03-08.
15. "School-Based Emergency preparedness : A National analysis and recommended Protocol" (PDF). Archive.ahrq.gov. Retrieved 2015-03-08.
16. AHRQ Publication No. 09-0013 January 2009 UMass System Office Hazard Mitigation Plan Draft (December 2013)
17. "Horse Farm Disaster Preparedness". TheHorse.com. 2014-11-26. Retrieved 2015-03-08.
18. Archived November 21, 2014, at the Wayback Machine.
19. "Your family's emergency kit is probably a disaster". Cnn.com. Retrieved 2015-03-08.
20. "Family Emergency Plan" (PDF). Ready.gov. Retrieved 2015-03-08.
21. "Home". Ready.gov. Retrieved 2015-03-08.
22. "Thunderclap: Resolve to be Ready in 2014!". Thunderclap.it. 2013-12-18. Retrieved 2015-03-08.
23. "Preparedness 101: Zombie Apocalypse | Public Health Matters Blog | Blogs | CDC". Blogs.cdc.gov. 2011-05-16. Retrieved 2015-03-08.
24. Archived December 17, 2014, at the Wayback Machine.
25. "Disaster Mental Health : Introduction". a1881.g.akamai.net. Retrieved 2015-03-08.
26. "Everyday Objects for Travel Disaster Survival". Travel Insurance Review. Retrieved 2015-03-08.
27. "Disaster Training". redcross.org. Retrieved 8 March 2015.
28. "Program Ops". Disaster Response Veterans Service Organization - Team Rubicon. Retrieved 8 March 2015.
29. "Why Prepare" (PDF). Fema.gov. Retrieved 2015-03-08.

30. "Basic Preparedness" (PDF). Fema.gov. Retrieved 2015-03-08.
31. "CDC Emergency preparedness and You | Gather Emergency Supplies | Disaster Supplies Kit". Emergency.cdc.gov. Retrieved 2015-03-08.
32. "Talking With Children" (PDF). Store.samhsa.gov. Retrieved 2015-03-08.
33. "Coping with Disaster". FEMA.gov. 2015-01-31. Retrieved 2015-03-08.
34. "Threat and Hazrd Identification and Risk Assessment Guide" (PDF). Fema.gov. August 2013. Retrieved 2015-03-08.
35. "My Patriot Supply : Homepage". Mypatriotsupply.com. Retrieved 2015-03-06.
36. "Murphy's Laws of TEOTWAWKI". Survivalcache.com. 2010-11-16. Retrieved 2015-03-08.
37. "American Preppers Network - National family survival and preparedness organization". American Preppers Network. Retrieved 8 March 2015.
38. "Man dies trying to rescue pets from fire". Pottsmerc.com. Retrieved 2015-03-08.
39. "Coping with Traumatic Events, SAMHSA.gov". Media.samhsa.gov. 2011-08-25. Retrieved 2015-03-08.
40. "Bug Out Bag Checklist". Theprepperproject.com. 2013-04-18. Retrieved 2015-03-08.
41. "Make A Plan". Ready.gov. 2014-01-29. Retrieved 2015-03-08.
42. "Escape From Fire!" (PDF). Usfa.fema.gov. Retrieved 2015-03-08.
43. "Who Will Qualify For Disability? - What Qualifying Is Based On". Ssdrc.com. Retrieved 2015-03-08.
44. "Preparing for Disaster for People with Disabilities and other Special Needs" (PDF). Redcross.org. Retrieved 2015-03-08.
45. (archived link, February 11, 2015)
46. "Ready New York : Preparing for Emergencies in new York City" (PDF). Nyc.com. Retrieved 2015-03-08.
47. "Install a Generator for Emergency Power" (PDF). Fema.gov. Retrieved 2015-03-08.
48. "Community Guidelines for Energy Emergencies | Department of Energy". Energy.gov. Retrieved 2015-03-08.
49. Generac Power Systems, Inc. "Generac Home Backup Power - Home & Portable Generator - Generac Power Systems". generac.com. Retrieved 8 March 2015.
50. "Duromax RV Grade 4,400-Watt 7.0 HP Gasoline Powered Portable Generator with Wheel Kit-XP4400 - The Home Depot". The Home Depot. 15 October 2014. Retrieved 8 March 2015.
51. "Microgrid Effects and Opportunities for Utilities" (PDF). Burnsmcd.com. Retrieved 2015-03-08.
52. "Alternative Energy Sources For Homes During Emergencies". House Hold Power Generator. 2013-08-18. Retrieved 2015-03-08.
53. "Goal Zero Yeti 400". Crutchfield. Retrieved 8 March 2015.

54. "Plan for Locations". Ready.gov. 2014-01-30. Retrieved 2015-03-08.
55. "Hospital Evacuation Decision Guide: Chapter 2. Pre-Disaster Self-Assessment". Archive.ahrq.gov. 2011-06-30. Retrieved 2015-03-08.
56. "Emergency Response Plan". Ready.gov. 2012-12-19. Retrieved 2015-03-08.
57. https://www.sba.gov/content/disaster-preparedness Emergency Preparedness
58. "FEMA.gov Communities - National Preparedness Community Main Group - Travel Preparedness Tips: How to Be Ready When On The Go". Community.fema.gov. 2014-07-31. Retrieved 2015-03-08.
59. "Commuter Emergency Plan" (PDF). Fema.gov. Retrieved 2015-03-06.
60. "Planes, Trains, and Automobiles - Holiday Travel Safety Tips". FEMA.gov. 2012-06-18. Retrieved 2015-03-08.
61. "Car Safety". Ready.gov. Retrieved 2015-03-08.
62. "Homeowner's Guide to Retrofitting" (PDF). Fema.gov. Retrieved 2015-03-08.
63. "Utility Shut-off & Safety". Ready.gov. Retrieved 2015-03-08.
64. "Federal Emergency Management Agency". FEMA.gov. Retrieved 2013-08-11.
65. Jaffin, Bob (September 17, 2008). "Emergency Management Training: How to Find the Right Program". Emergency Management Magazine. Retrieved 2008-11-15.
66. "Certification-General CEM certification Info". Iaem.com. Retrieved 2012-03-07.
67. "Principles of Emergency Management Supplement" (PDF). 2007-09-11. Retrieved 2015-03-06.
68. "Smart Emergency Response System (SERS)". Smart America. Retrieved 2015-03-08.
69. "The Smart Emergency Response System Using MATLAB and Simulink". YouTube. 2014-07-18. Retrieved 2015-03-08.
70. Buchanan, Sally. "Emergency preparedness." from Paul Banks and Roberta Pilette. Preservation Issues and Planning. Chicago: American Library Association, 2000. 159–165. ISBN 978-0-8389-0776-4
71. "FEWS Network — USAID". Population Explorer. Retrieved 2012-03-07.
72. The Veterinary profession's duty of care in response to disasters and food animal emergencies. Journal of the American Veterinary Medical Association, Vol 231, No. 2, July 15, 2007
73. "IAEMAsia". Iaem.com. Retrieved 2012-03-07.
74. "Region 13". Iaem.com. Retrieved 2012-03-07.
75. "IAEM Europa". Iaem.com. Retrieved 2012-03-07.
76. "IAEM Oceania". Iaem.com. Retrieved 2012-03-07.
77. "Welcome to IAEM.COM- International Association of Emergency Managers". Iaem.com. Retrieved 2012-03-07.
78. "Iaem-Usa". Iaem.com. Retrieved 2012-03-07.

79. "Welcome to the International Recovery Platform — International Recovery Platform". Recoveryplatform.org. Retrieved 2015-01-15.

80. "Field Assessment Coordination Teams (FACT)". IFRC. 2011-10-15. Retrieved 2012-03-07.

81. "Emergency Response Units (ERUs)". IFRC. 2011-10-15. Retrieved 2012-03-07.

82. https://gobgr.org/about/about

83. "Disaster Risk Management - Projects". Web.worldbank.org. 2004-04-28. Retrieved 2015-03-08.

84. "Core Data Sets". Ldeo.columbia.edu. Retrieved 2015-03-08.

85. https://web.archive.org/web/20080423231412/http://gfdrr.org/index.cfm?Page=home&ItemID=200. Archived from the original on April 23, 2008. Retrieved March 25, 2008. Missing or empty |title= (help)

86. Boin, A.; Rhinard, M. (2008). "Managing Transboundary Crises: What Role for the European Union?". International Studies Review. 10: 1–26. doi:10.1111/j.1468-2486.2008.00745.x.

87. "Civil Protection – The Community mechanism for civil protection". Ec.europa.eu. Retrieved 2010-07-29.

88. Marc Jansen (2010-06-29). "Startseite des Studiengangs Katastrophenvorsorge und -management". Kavoma.de. Retrieved 2010-07-29.

89. "Functions and Responsibilities". National Disaster Management Authority. Retrieved 2014-10-28.

90. "About Us". National Disaster Response Force. Retrieved 2015-01-15.

91. National Civil Defence Emergency Plan Order 2005. Legislation.govt.nz (2008-10-01). Retrieved on 2011-07-28.

92. "Civil Defence". beehive.govt.nz. Retrieved 2015-03-08.

93. Archived August 10, 2008, at the Wayback Machine.

94. "Civil Defence Emergency Management Act 2002 No 33 (as at 01 January 2014), Public Act 4 Interpretation – New Zealand Legislation". Legislation.govt.nz. Retrieved 2015-03-08.

95. For example, disaster is not used in the Civil Defence Emergency Management Act 2002, the enabling legislation for New Zealand's emergency management

96. Retrieved 3 August according to rahul jain the fludes and natural uncertainties are included in mgt it is known as disaster mgt2008. (PDF) . Retrieved on 2011-07-28.

97. National Civil Defence Emergency Management Strategy 2007, page 5. Department of Internal Affairs, Wellington, New Zealand 2008. Digital edition. Retrieved 3 August 2008. ISBN 0-478-29453-0.

98. Prime Minister's Media Office (30 May 2013). "SOMALIA: Prime Minister calls for parliament to enact legislation as Cabinet moves to establish disaster management agency". Raxanreeb. Retrieved 5 June 2013.

99. "Welcome - The Emergency Planning Society". The-eps.org. Retrieved 2015-03-08.
100. "Institute of Civil Protection & Emergency Management | Welcome". ICPEM. 2014-04-03. Retrieved 2015-03-08.
101. "UK | Northern Ireland | Mock plane crash tests NI crews". BBC News. 2007-05-20. Retrieved 2015-03-08.
102. McElreath, David; Doss, Daniel; Jensen, Carl; Wigginton, Michael; Nations, Robert; Van Slyke, Jeffrey; Nations, Julie (2014). Foundations of Emergency Management (1st ed.). Dubuque, IA: Kendall-Hunt Publishing Company. p. 25. ISBN 978-1465234889.
103. Doss, Daniel; Glover, William; Goz, Rebecca; Wigginton, Michael (2015). The Foundations of Communication in Criminal Justice Systems. Boca Raton, Florida: CRC Press. p. 301. ISBN 978-1482236576.
104. Doss, Daniel; Glover, William; Goza, Rebecca; Wigginton, Michael (2015). The Foundations of Communication in Criminal Justice Systems (1 ed.). Boca Raton, Florida: CRC Press. p. 301. ISBN 978-1482236576.
105. "The Salvation Army Emergency Disaster Services". Disaster.salvationarmyusa.org. Retrieved 2015-03-08.
106. "2012 Activity Report". Namb.net. Retrieved 2015-03-08.
107. http://www.baptistrelief.org/Our_Work/
108. "Citizen Corps | Ready.gov" (PDF). Citizencorps.gov. 2013-07-23. Retrieved 2013-08-11.
109. http://training.fema.gov/
110. EMI Accredited
111. EMI Overview
112. FEMA EMI Independent Study home
113. IS Course Offerings
114. "Emergency Response | Catholic Relief Services". Crs.org. Retrieved 2015-03-08.

CHAPTER 9

Geotechnical Engineering

Chapter Contents

- **Introduction**
- **History**
- **Practicing Engineers**
- **Soil Mechanics**
- **Geotechnical Investigation**
- **Foundations**
- **Lateral Earth Support Structures**
- **Earthworks**
- **Ground Improvement**
- **Slope Stabilization**
- **Offshore Geotechnical Engineering**
- **Geosynthetics**
- **Soil Liquefaction**
- **Response Spectrum**
- **Seismic Hazard**
- **Mces**
- **Ground–Structure Interaction**

From a scientific perspective, geotechnical engineering largely involves defining the soil's strength and deformation properties. Clay, silt, sand, rock and snow are important materials in geotechnics. Geotechnical engineering includes specialist fields such as soil and rock mechanics, geophysics, hydrogeology and associated disciplines such as geology.

INTRODUCTION

Geotechnical engineering is the branch of civil engineering concerned with the engineering behavior of earth materials. Geotechnical engineering is important in civil engineering, but also has applications in military, mining, petroleum and other engineering disciplines that are concerned with construction occurring on the surface or within the ground. Geotechnical engineering uses principles of soil mechanics and rock mechanics to investigate subsurface conditions and materials; determine the relevant physical/mechanical and chemical properties of these materials; evaluate stability of natural slopes and man-made soil deposits; assess risks posed by site conditions; design earthworks and structure foundations; and monitor site conditions, earthwork and foundation construction.

A typical geotechnical engineering project begins with a review of project needs to define the required material properties. Then follows a site investigation of soil, rock, fault distribution and bedrock properties on and below an area of interest to determine their engineering properties including how they will interact with, on or in a proposed construction. Site investigations are needed to gain an understanding of the area in or on which the engineering will take place. Investigations can include the assessment of the risk to humans, property and the environment from natural hazards such as earthquakes, landslides, sinkholes, soil liquefaction, debris flows and rockfalls.

Boston's Big Dig presented geotechnical challenges in an urban environment.

A geotechnical engineer then determines and designs the type of foundations, earthworks, and/or pavement subgrades required for the intended man-made structures to be built. Foundations are designed and constructed for structures of various sizes such as high-rise buildings, bridges, medium to large commercial buildings, and smaller structures where the soil conditions do not allow code-based design.

Foundations built for above-ground structures include shallow and deep foundations. Retaining structures include earth-filled dams and retaining walls. Earthworks include embankments, tunnels, dikes and levees, channels, reservoirs, deposition of hazardous waste and sanitary landfills.

Geotechnical engineering is also related to coastal and ocean engineering. Coastal engineering can involve the design and construction of wharves, marinas, and jetties. Ocean engineering can involve foundation and anchor systems for offshore structures such as oil platforms.

The fields of geotechnical engineering and engineering geology are closely related, and have large areas of overlap. However, the field of geotechnical engineering is a specialty of engineering, where the field of engineering geology is a specialty of geology.

HISTORY

Humans have historically used soil as a material for flood control, irrigation purposes, burial sites, building foundations, and as construction material for buildings. First activities were linked to irrigation and flood control, as demonstrated by traces of dykes, dams, and canals dating back to at least 2000 BCE that were found in ancient Egypt, ancient Mesopotamia and the Fertile Crescent, as well as around the early settlements of Mohenjo Daro and Harappa in the Indus valley. As the cities expanded, structures were erected supported by formalized foundations; Ancient Greeks notably constructed pad footings and strip-and-raft foundations. Until the 18th century, however, no theoretical basis for soil design had been developed and the discipline was more of an art than a science, relying on past experience.

Several foundation-related engineering problems, such as the Leaning Tower of Pisa, prompted scientists to begin taking a more scientific-based approach to examining the subsurface. The earliest advances occurred in the development of earth pressure theories for the construction of retaining walls. Henri Gautier, a French Royal Engineer, recognized the "natural slope" of different soils in 1717, an idea later known as the soil's angle of repose. A rudimentary soil classification system was also developed based on a material's unit weight, which is no longer considered a good indication of soil type.

The application of the principles of mechanics to soils was documented as early as 1773 when Charles Coulomb (a physicist, engineer, and army Captain) developed improved methods to determine the earth pressures against military ramparts. Coulomb observed that, at failure, a distinct slip plane would form behind a sliding retaining wall and he suggested that the maximum shear stress on the slip plane, for design purposes, was the sum of the soil cohesion, c, and friction $\sigma \tan(\phi)$, where σ is the normal stress on the slip plane and ϕ is the friction angle of the soil. By combining Coulomb's theory with Christian Otto Mohr's 2D stress state, the theory became known as Mohr-Coulomb theory. Although it is now recognized that precise determination of cohesion is impossible because c is not a fundamental soil property, the Mohr-Coulomb theory is still used in practice today.

In the 19th century Henry Darcy developed what is now known as Darcy's Law describing the flow of fluids in porous media. Joseph Boussinesq (a mathematician and physicist) developed theories of stress distribution in elastic solids that proved useful for estimating stresses at depth in the ground; William Rankine, an engineer and physicist, developed an alternative to Coulomb's earth pressure theory. Albert Atterberg developed the clay consistency indices that are still used today for soil classification. Osborne Reynolds recognized in 1885 that shearing causes volumetric dilation of dense and contraction of loose granular materials.

Modern geotechnical engineering is said to have begun in 1925 with the publication of *Erdbaumechanik* by Karl Terzaghi (a civil engineer and geologist). Considered by many to be the father of modern soil mechanics and geotechnical engineering, Terzaghi developed the principle of effective stress, and demonstrated that the shear strength of soil is controlled by effective stress. Terzaghi also developed the framework for theories of bearing capacity of foundations, and the theory for prediction of the rate of settlement of clay layers due to consolidation. In his 1948 book, Donald Taylor recognized that interlocking and dilation of densely packed particles contributed to the peak strength of a soil. The interrelationships between volume change behavior (dilation, contraction, and consolidation) and shearing behavior were all connected via the theory of plasticity using critical state soil mechanics by Roscoe, Schofield, and Wroth with the publication of "On the Yielding of Soils" in 1958. Critical state soil mechanics is the basis for many contemporary advanced constitutive models describing the behavior of soil.

Geotechnical centrifuge modeling is a method of testing physical scale models of geotechnical problems. The use of a centrifuge enhances the similarity of the scale model tests involving soil because the strength and stiffness of soil is very sensitive to the confining pressure. The centrifugal acceleration allows a researcher to obtain large (prototype-scale) stresses in small physical models.

PRACTICING ENGINEERS

Geotechnical engineers are typically graduates of a four-year civil engineering program and some hold a masters degree. In the USA, geotechnical engineers are typically licensed and regulated as Professional Engineers (PEs) in most states; currently only California and Oregon have licensed geotechnical engineering specialties. The Academy of Geo-Professionals (AGP) began issuing Diplomate, Geotechnical Engineering (D.GE) certification in 2008. State governments will typically license engineers who have graduated from an ABET accredited school, passed the Fundamentals of Engineering examination, completed several years of work experience under the supervision of a licensed Professional Engineer, and passed the Professional Engineering examination.

SOIL MECHANICS

In geotechnical engineering, soils are considered a three-phase material composed of: rock or mineral particles, water and air. The voids of a soil, the spaces in between mineral particles, contain the water and air.

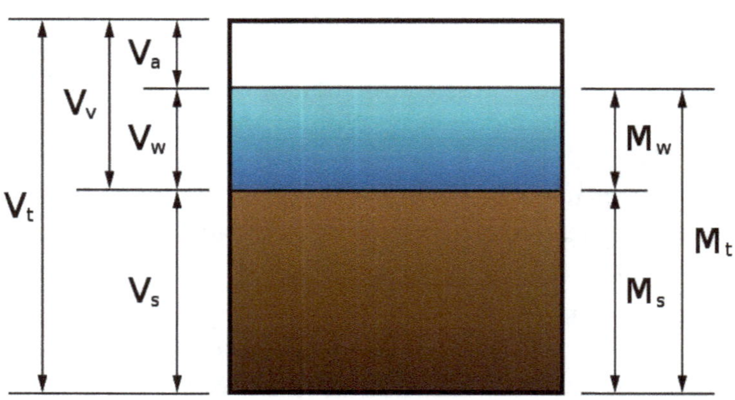

A phase diagram of soil indicating the weights and volumes of air, soil, water, and voids.

The engineering properties of soils are affected by four main factors: the predominant size of the mineral particles, the type of mineral particles, the grain size distribution, and the relative quantities of mineral, water and air present in the soil matrix. Fine particles (fines) are defined as particles less than 0.075 mm in diameter.

Soil Properties

Some of the important properties of soils that are used by geotechnical engineers to analyze site conditions and design earthworks, retaining structures, and foundations are:

Specific Weight or Unit Weight

Cumulative weight of the solid particles, water and air of the unit volume of soil. Note that the air phase is often assumed to be weightless.

Porosity

Ratio of the volume of voids (containing air, water, or other fluids) in a soil to the total volume of the soil. Porosity is mathematically related to void ratio the by

$$n = \frac{e}{1+e}$$

here e is void ratio and n is porosity

Void Ratio

The ratio of the volume of voids to the volume of solid particles in a soil mass. Void ratio is mathematically related to the porosity by

$$e = \frac{n}{1-n}$$

Permeability

A measure of the ability of water to flow through the soil. It is expressed in units of velocity.

Compressibility

The rate of change of volume with effective stress. If the pores are filled with water, then the water must be squeezed out of the pores to allow volumetric compression of the soil; this process is called consolidation.

Shear Strength

The maximum shear stress that can be applied in a soil mass without causing shear failure.

Atterberg Limits

Liquid limit, Plastic limit, and Shrinkage limit. These indices are used for estimation of other engineering properties and for soil classification.

GEOTECHNICAL INVESTIGATION

Geotechnical engineers and engineering geologists perform geotechnical investigations to obtain information on the physical properties of soil and rock underlying (and sometimes adjacent to) a site to design earthworks and foundations for proposed structures, and for repair of distress to earthworks and structures caused by subsurface conditions. A geotechnical investigation will include surface exploration and subsurface exploration of a site. Sometimes, geophysical methods are used to obtain data about sites. Subsurface exploration usually involves in-situ testing (two common examples of in-situ tests are the standard penetration test and cone penetration test). In addition site investigation will often include subsurface sampling and laboratory testing of the soil samples retrieved. The digging of test pits and trenching (particularly for locating faults and slide planes) may also be used to learn about soil conditions at depth. Large diameter borings are rarely used due to safety concerns and expense, but are sometimes used to allow a geologist or engineer to be lowered into the borehole for direct visual and manual examination of the soil and rock stratigraphy.

A variety of soil samplers exist to meet the needs of different engineering projects. The standard penetration test (SPT), which uses a thick-walled split spoon sampler, is the most common way to collect disturbed samples. Piston samplers, employing a thin-walled tube, are most commonly used for the collection of less disturbed samples. More advanced methods, such as ground freezing and the Sherbrooke block sampler, are superior, but even more expensive.

Atterberg limits tests, water content measurements, and grain size analysis, for example, may be performed on disturbed samples obtained from thick walled soil samplers. Properties such as shear strength, stiffness hydraulic conductivity, and coefficient of consolidation may be significantly altered by sample disturbance. To measure these properties in the laboratory, high quality sampling is required. Common tests to measure the strength and stiffness include the triaxial shear and unconfined compression test.

Surface exploration can include geologic mapping, geophysical methods, and photogrammetry; or it can be as simple as an engineer walking around to observe the physical conditions at the site. Geologic mapping and interpretation of geomorphology is typically completed in consultation with a geologist or engineering geologist. Geophysical exploration is also sometimes used. Geophysical techniques used for subsurface exploration include measurement of seismic waves (pressure, shear, and Rayleigh waves), surface-wave methods and/or downhole methods, and electromagnetic surveys (magnetometer, resistivity, and ground-penetrating radar).

FOUNDATIONS

A building's foundation transmits loads from buildings and other structures to the earth. Geotechnical engineers design foundations based on the load characteristics of the structure and the properties of the soils and/or bedrock at the site. In general, geotechnical engineers:

- Estimate the magnitude and location of the loads to be supported.
- Develop an investigation plan to explore the subsurface.
- Determine necessary soil parameters through field and lab testing (e.g., consolidation test, triaxial shear test, vane shear test, standard penetration test).
- Design the foundation in the safest and most economical manner.

The primary considerations for foundation support are bearing capacity, settlement, and ground movement beneath the foundations. Bearing capacity is the ability of the site soils to support the loads imposed by buildings or structures. Settlement occurs under all foundations in all soil conditions, though lightly loaded structures or rock sites may experience negligible settlements. For heavier structures or softer sites, both overall settlement relative to unbuilt areas or neighboring buildings, and differential settlement under a single structure, can be concerns. Of particular concern is settlement which occurs over time, as immediate settlement can usually be compensated for during construction. Ground movement beneath a structure's foundations can occur due to shrinkage or swell of expansive soils due to climatic changes, frost expansion of soil, melting of permafrost, slope instability, or other causes. All these factors must be considered during design of foundations.

Many building codes specify basic foundation design parameters for simple conditions, frequently varying by jurisdiction, but such design techniques are normally limited to certain types of construction and certain types of sites, and are frequently very conservative.

In areas of shallow bedrock, most foundations may bear directly on bedrock; in other areas, the soil may provide sufficient strength for the support of structures. In areas of deeper bedrock with soft overlying soils, deep foundations are used to support structures directly on the bedrock; in areas where bedrock is not economically available, stiff "bearing layers" are used to support deep foundations instead.

Shallow Foundations

Shallow foundations are a type of foundation that transfers building load to the very near the surface, rather than to a subsurface layer. Shallow foundations typically have a depth to width ratio of less than 1.

Example of a slab-on-grade foundation.

Footings

Footings (often called "spread footings" because they spread the load) are structural elements which transfer structure loads to the ground by direct areal contact. Footings can be isolated footings for point or column loads, or strip footings for wall or other long (line) loads. Footings are normally constructed from reinforced concrete cast directly onto the soil, and are typically embedded into the ground to penetrate through the zone of frost movement and/or to obtain additional bearing capacity.

Slab foundations

A variant on spread footings is to have the entire structure bear on a single slab of concrete underlying the entire area of the structure. Slabs must be thick enough to provide sufficient rigidity to spread the bearing loads somewhat uniformly, and to minimize differential settlement across the foundation. In some cases, flexure is allowed and the building is constructed to tolerate small movements of the foundation instead. For small structures, like single-family houses, the slab may be less than 300 mm thick; for larger structures, the foundation slab may be several meters thick.

Slab foundations can be either slab-on-grade foundations or embedded foundations, typically in buildings with basements. Slab-on-grade foundations must be designed to allow for potential ground movement due to changing soil conditions.

Deep Foundations

Deep foundations are used for structures or heavy loads when shallow foundations cannot provide adequate capacity, due to size and structural limitations. They may also be used to transfer building loads past weak or compressible soil layers. While shallow foundations rely solely on the bearing capacity of the soil beneath them, deep foundations can rely on end bearing resistance, frictional resistance along their length, or both in developing the required capacity. Geotechnical engineers use specialized tools, such as the cone penetration test, to estimate the amount of skin and end bearing resistance available in the subsurface.

Pile-driving for a bridge in Napa, California.

There are many types of deep foundations including piles, drilled shafts, caissons, piers, and earth stabilized columns. Large buildings such as skyscrapers typically require deep foundations. For example, the Jin Mao Tower in China uses tubular steel piles about 1m (3.3 feet) driven to a depth of 83.5m (274 feet) to support its weight.

In buildings that are constructed and found to undergo settlement, underpinning piles can be used to stabilise the existing building.

There are three ways to place piles for a deep foundation. They can be driven, drilled, or installed by use of an auger. Driven piles are extended to their necessary depths with the application of external energy in the same way a nail is hammered. There are four typical hammers used to drive such piles: drop hammers, diesel hammers, hydraulic hammers, and air hammers. Drop hammers simply drop a heavy weight onto the pile to drive it, while diesel hammers use a single cylinder diesel engine to force piles through the Earth. Similarly, hydraulic and air hammers supply energy to piles through hydraulic and air forces. Energy imparted from a hammer head varies with type of hammer chosen, and can be as high as a million foot pounds for large scale diesel hammers, a very common hammer head used in practice. Piles are made of a variety of material including steel, timber, and concrete. Drilled piles are created by first drilling a hole to the appropriate depth, and filling it with concrete. Drilled piles can typically carry more load than driven piles, simply due to a larger diameter pile. The auger method of pile installation is similar to drilled pile installation, but concrete is pumped into the hole as the auger is being removed.

LATERAL EARTH SUPPORT STRUCTURES

A retaining wall is a structure that holds back earth. Retaining walls stabilize soil and rock from downslope movement or erosion and provide support for vertical or near-vertical grade changes. Cofferdams and bulkheads, structures to hold back water, are sometimes also considered retaining walls.

The primary geotechnical concern in design and installation of retaining walls is that the weight of the retained material is creates lateral earth pressure behind the wall, which can cause the wall to deform or fail. The lateral earth pressure depends on the height of the wall, the density of the soil,the strength of the soil, and the amount of allowable movement of the wall. This pressure is smallest at the top and increases toward the bottom in a manner similar to hydraulic pressure, and tends to push the wall away from the backfill. Groundwater behind the wall that is not dissipated by a drainage system causes an additional horizontal hydraulic pressure on the wall.

Gravity Walls

Gravity walls depend on the size and weight of the wall mass to resist pressures from behind. Gravity walls will often have a slight setback, or batter, to improve wall stability. For short, landscaping walls, gravity walls made from dry-stacked (mortarless) stone or segmental concrete units (masonry units) are commonly used.

Earlier in the 20th century, taller retaining walls were often gravity walls made from large masses of concrete or stone. Today, taller retaining walls are increasingly built as composite gravity walls such as: geosynthetic or steel-reinforced backfill soil with precast facing; gabions (stacked steel wire baskets filled with rocks), crib walls (cells built up log cabin style from precast concrete or timber and filled with soil or free draining gravel) or soil-nailed walls (soil reinforced in place with steel and concrete rods).

For reinforced-soil gravity walls, the soil reinforcement is placed in horizontal layers throughout the height of the wall. Commonly, the soil reinforcement is geogrid, a high-strength polymer mesh, that provide tensile strength to hold soil together. The wall face is often of precast, segmental concrete units that can tolerate some differential movement. The reinforced soil's mass, along with the facing, becomes the gravity wall. The reinforced mass must be built large enough to retain the pressures from the soil behind it. Gravity walls usually must be a minimum of 30 to 40 percent as deep (thick) as the height of the wall, and may have to be larger if there is a slope or surcharge on the wall.

Cantilever Walls

Prior to the introduction of modern reinforced-soil gravity walls, cantilevered walls were the most common type of taller retaining wall. Cantilevered walls are made from a relatively thin stem of steel-reinforced, cast-in-place concrete or mortared masonry (often in the shape of an inverted T). These walls cantilever loads (like a beam) to a large, structural footing; converting horizontal pressures from behind the wall to vertical pressures on the ground below. Sometimes cantilevered walls are buttressed on the front, or include a counterfort on the back, to improve their stability against high loads. Buttresses are short wing walls at right angles to the main trend of the wall. These walls require rigid concrete footings below seasonal frost depth. This type of wall uses much less material than a traditional gravity wall.

Cantilever walls resist lateral pressures by friction at the base of the wall and/or **passive earth pressure**, the tendency of the soil to resist lateral movement.

Basements are a form of cantilever walls, but the forces on the basement walls are greater than on conventional walls because the basement wall is not free to move.

Excavation Shoring

Shoring of temporary excavations frequently requires a wall design which does not extend laterally beyond the wall, so shoring extends below the planned base of the excavation. Common methods of shoring are the use of sheet piles or soldier beams and lagging. Sheet piles are a form of driven piling using thin interlocking sheets of steel to obtain a continuous barrier in the ground, and are driven prior to excavation. Soldier beams are constructed of wide flange steel H sections spaced about 2–3 m apart, driven prior to excavation. As the excavation proceeds, horizontal timber or steel sheeting (lagging) is inserted behind the H pile flanges

In some cases, the lateral support which can be provided by the shoring wall alone is insufficient to resist the planned lateral loads; in this case additional support is provided by walers or tie-backs. Walers are structural elements which connect across the excavation so that the loads from the soil on either side of the excavation are used to resist each other, or which transfer horizontal loads from the shoring wall to the base of the excavation. Tie-backs are steel tendons drilled into the face of the wall which extend beyond the soil which is applying pressure to the wall, to provide additional lateral resistance to the wall.

EARTHWORKS

Excavation

Excavation is the process of training earth according to requirement by removing the soil from the site.

Filling

Filling is the process of training earth according to requirement by placing the soil on the site.

Compaction

Compaction is the process by which the density of soil is increased and permeability of soil is decreased. Fill placement work often has specifications requiring a specific degree of compaction, or alternatively, specific properties of the compacted soil. In-situ soils can be compacted by rolling, deep dynamic compaction, vibration, blasting, gyrating, kneading, compaction grouting etc.

A compactor/roller operated by U.S. Navy Seabees

GROUND IMPROVEMENT

Ground Improvement is a technique that improves the engineering properties of the treated soil mass. Usually, the properties modified are shear strength, stiffness and permeability. Ground improvement has developed into a sophisticated tool to support foundations for a wide variety of structures. Properly applied, i.e. after giving due consideration to the nature of the ground being improved and the type and sensitivity of the structures being built, ground improvement often reduces direct costs and saves time.

SLOPE STABILIZATION

Slope stability is the potential of soil covered slopes to withstand and undergo movement. Stability is determined by the balance of shear stress and shear strength. A previously stable slope may be initially affected by preparatory factors, making the slope conditionally unstable. Triggering factors of a slope failure can be climatic events can then make a slope actively unstable, leading to mass movements. Mass movements can be caused by increases in shear stress, such as loading, lateral pressure, and transient forces. Alternatively, shear strength may be decreased by weathering, changes in pore water pressure, and organic material.

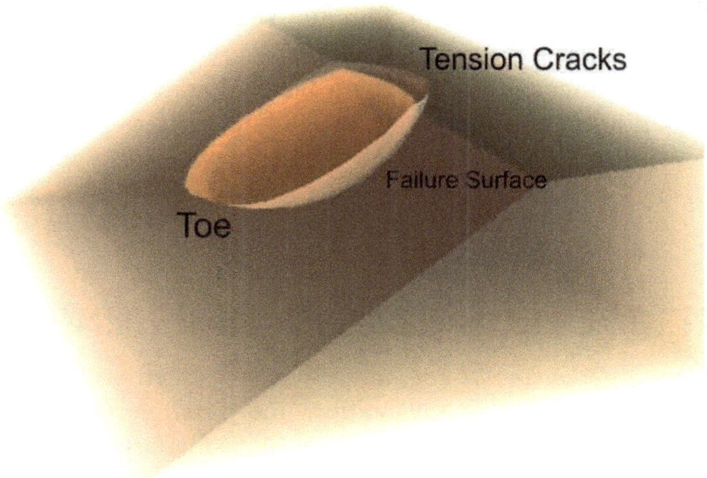

Simple slope slip section.

Several modes of failure for earth slopes include falls, topples, slides, and flows. In slopes with coarse grained soil or rocks, falls typically occur as the rapid descent of rocks and other loose slope material. A slope topples when a large column of soil tilts over its vertical axis at failure. Typical slope stability analysis considers sliding failures, categorized mainly as rotational slides or translational slides. As implied by the name, rotational slides fail along a generally curved surface, while translational slides fail along a more planar surface. A slope failing as a flow would resemble a fluid flowing downhill.

Slope Stability Analysis

Stability analysis is needed for the design of engineered slopes and for estimating the risk of slope failure in natural or designed slopes. A common assumption is that a slope consists of a layer of soil sitting on top of a rigid base. The mass and the base are assumed to interact via friction. The interface between the mass and the base can be planar, curved, or have some other complex geometry. The goal of a slope stability analysis is to determine the conditions under which the mass will slip relative to the base and lead to slope failure.

If the interface between the mass and the base of a slope has a complex geometry, slope stability analysis is difficult and numerical solution methods are required. Typically, the exact geometry of the interface is not known and a simplified interface geometry is assumed. Finite slopes require three-dimensional models to be analyzed. To keep the problem simple, most slopes are analyzed assuming that the slopes are infinitely wide and can therefore be represented by two-dimensional models. A slope can be drained or undrained. The undrained condition is used in the calculations to produce conservative estimates of risk.

A popular stability analysis approach is based on principles pertaining to the limit equilibrium concept. This method analyzes a finite or infinite slope as if it were about to fail along its sliding failure surface. Equilibrium stresses are calculated along the failure plane, and compared to the soils shear strength as determined by Terzaghi's shear strength equation. Stability is ultimately decided by a factor of safety equal to the ratio of shear strength to the equilibrium stresses along the failure surface. A factor of safety greater than one generally implies a stable slope, failure of which should not occur assuming the slope is undisturbed. A factor of safety of 1.5 for static conditions is commonly used in practice.

OFFSHORE GEOTECHNICAL ENGINEERING

Offshore (or *marine*) *geotechnical engineering* is concerned with foundation design for human-made structures in the sea, away from the coastline (in opposition to *onshore* or *nearshore*). Oil platforms, artificial islands and submarine pipelines are examples of such structures. There are number of significant differences between onshore and offshore geotechnical engineering. Notably, ground improvement (on the seabed) and site investigation are more expensive, the offshore structures are exposed to a wider range of geohazards, and the environmental and financial consequences are higher in case of failure. Offshore structures are exposed to various environmental loads, notably wind, waves and currents. These phenomena may affect the integrity or the serviceability of the structure and its foundation during its operational lifespan – they need to be taken into account in offshore design.

Platforms offshore Mexico.

In subsea geotechnical engineering, seabed materials are considered a two-phase material composed of 1) rock or mineral particles and 2) water. Structures may be fixed in place in the seabed—as is the case for piers, jettys and fixed-bottom wind turbines—or may be a floating structure that remain roughly fixed relative to its geotechnical anchor point. Undersea mooring of human-engineered floating structures include a large number of offshore oil and gas platforms and, since 2008, a few floating wind turbines. Two common types of engineered design for anchoring floating structures include tension-leg and catenary loose mooring systems. "Tension leg mooring systems have vertical tethers under tension providing large restoring moments in pitch and roll. Catenary mooring systems provide station keeping for an offshore structure yet provide little stiffness at low tensions."

GEOSYNTHETICS

Geosynthetics are a type of plastic polymer products used in geotechnical engineering that improve engineering performance while reducing costs. This includes geotextiles, geogrids, geomembranes, geocells, and geocomposites.

A collage of geosynthetic products.

The synthetic nature of the products make them suitable for use in the ground where high levels of durability are required; their main functions include: drainage, filtration, reinforcement, separation and containment. Geosynthetics are available in a wide range of forms and materials, each to suit a slightly different end use, although they are frequently used together. These products have a wide range of applications and are currently used in many civil and geotechnical engineering applications including: roads, airfields, railroads, embankments, piled embankments, retaining structures, reservoirs, canals, dams, landfills, bank protection and coastal engineering.

SOIL LIQUEFACTION

Soil liquefaction describes a phenomenon whereby a saturated or partially saturated soil substantially loses strength and stiffness in response to an applied stress, usually earthquake shaking or other sudden change in stress condition, causing it to behave like a liquid.

In soil mechanics the term "liquefied" was first used by Allen Hazen in reference to the 1918 failure of the Calaveras Dam in California. He described the mechanism of flow liquefaction of the embankment dam as follows:

If the pressure of the water in the pores is great enough to carry all the load, it will have the effect of holding the particles apart and of producing a condition that is practically equivalent to that of quicksand... the initial movement of some part of the material might result in accumulating pressure, first on one point, and then on another, successively, as the early points of concentration were liquefied.

Some effects of liquefaction after the 1964 Niigata earthquake

Liquefaction allowed this sewer to float upward – 2004 Chūetsu earthquake

The phenomenon is most often observed in saturated, loose (low density or uncompacted), sandy soils. This is because a loose sand has a tendency to compress when a load is applied; dense sands by contrast tend to expand in volume or 'dilate'. If the soil is saturated by water, a condition that often exists when the soil is below the ground water table or sea level, then water fills the gaps between soil grains ('pore spaces'). In response to the soil compressing, this water increases in pressure and attempts to flow out from the soil to zones of low pressure (usually upward towards the ground surface). However, if the loading is rapidly applied and large enough, or is repeated many times (e.g. earthquake shaking, storm wave loading) such that it does not flow out in time before the next cycle of load is applied, the water pressures may build to an extent where they exceed the contact stresses between the grains of soil that keep them in contact with each other. These contacts between grains are the means by which the weight from buildings and overlying soil layers are transferred from the ground surface to layers of soil or rock at greater depths. This loss of soil structure causes it to lose all of its strength (the ability to transfer shear stress) and it may be observed to flow like a liquid (hence 'liquefaction').

The effect of liquefaction in Christchurch, New Zealand, during the Mw 6.3 February 2011 Christchurch earthquake

Although the effects of liquefaction have been long understood, it was more thoroughly brought to the attention of engineers after the 1964 Niigata earthquake and 1964 Alaska earthquake. It was also a major factor in the destruction in San Francisco's Marina District during the 1989 Loma Prieta earthquake, and in Port of Kobe during the 1995 Great Hanshin earthquake. More recently liquefaction was largely responsible for extensive damage to residential properties in the eastern suburbs and satellite townships of Christchurch, New Zealand during the 2010 Canterbury earthquake and more extensively again following the Christchurch earthquakes that followed in early and mid-2011. The building codes in many developed countries require engineers to consider the effects of soil liquefaction in the design of new buildings and infrastructure such as bridges, embankment dams and retaining structures.

Technical Definitions

A state of 'soil liquefaction' occurs when the effective stress of soil is reduced to essentially zero, which corresponds to a complete loss of shear strength. This may be initiated by either monotonic loading (e.g. single sudden occurrence of a change in stress – examples include an increase in load on an embankment or sudden loss of toe support) or cyclic loading (e.g. repeated change in stress condition – examples include wave loading or earthquake shaking). In both cases a soil in a saturated loose state, and one which may generate significant pore water pressure on a change in load are the most likely to liquefy. This is because a loose soil has the tendency to compress when sheared, generating large excess porewater pressure as load is transferred from the soil skeleton to adjacent pore water during undrained loading. As pore water pressure rises a progressive loss of strength of the soil occurs as effective stress is reduced. It is more likely to occur in sandy or non-plastic silty soils, but may in rare cases occur in gravels and clays.

A 'flow failure' may initiate if the strength of the soil is reduced below the stresses required to maintain equilibrium of a slope or footing of a building for instance. This can occur due to monotonic loading or cyclic loading, and can be sudden and catastrophic. A historical example is the Aberfan disaster. Casagrande referred to this type of phenomena as 'flow liquefaction' although a state of zero effective stress is not required for this to occur.

The term 'cyclic liquefaction' refers to the occurrence of a state of soil when large shear strains have accumulated in response to cyclic loading. A typical reference strain for the approximate occurrence of zero effective stress is 5% double amplitude shear strain. This is a soil test based definition, usually performed via cyclic triaxial, cyclic direct simple shear, or cyclic torsional shear type apparatus. These tests are performed to determine a soil's resistance to liquefaction by observing the number of cycles of loading at a particular shear stress amplitude before it 'fails'. Failure here is defined by the aforementioned shear strain criteria.

The term 'cyclic mobility' refers to the mechanism of progressive reduction of effective stress due to cyclic loading. This may occur in all soil types including dense soils. However, on reaching a state of zero effective stress such soils immediate dilate and regain strength. Thus shear strains are significantly less than a true state of soil liquefaction whereby a loose soil exhibits flow type phenomena.

Occurrence

Liquefaction is more likely to occur in loose to moderately saturated granular soils with poor drainage, such as silty sands or sands and gravels capped or containing seams of impermeable sediments. During wave loading, usually cyclic undrained loading, e.g. seismic loading, loose sands tend to decrease in volume, which produces an increase in their pore water pressures and consequently a decrease in shear strength, i.e. reduction in effective stress.

Deposits most susceptible to liquefaction are young (Holocene-age, deposited within the last 10,000 years) sands and silts of similar grain size (well-sorted), in beds at least metres thick, and saturated with water. Such deposits are often found along stream beds, beaches, dunes, and areas where windblown silt (loess) and sand have accumulated. Some examples of soil liquefaction include quicksand, quick clay, turbidity currents, and earthquake induced liquefaction.

Depending on the initial void ratio, the soil material can respond to loading either *strain-softening* or *strain-hardening*. Strain-softened soils, e.g. loose sands, can be triggered to collapse, either monotonically or cyclically, if the static shear stress is greater than the ultimate or steady-state shear strength of the soil. In this case *flow liquefaction* occurs, where the soil deforms at a low constant residual shear stress. If the soil strain-hardens, e.g. moderately dense to dense sand, flow liquefaction will generally not occur. However, *cyclic softening* can occur due to cyclic undrained loading, e.g. earthquake loading. Deformation during cyclic loading will depend on the density of the soil, the magnitude and duration of the cyclic loading, and amount of shear stress reversal. If stress reversal occurs, the effective shear stress could reach zero, then *cyclic liquefaction* can take place. If stress reversal does not occur, zero effective stress is not possible to occur, then *cyclic mobility* takes place.

The resistance of the cohesionless soil to liquefaction will depend on the density of the soil, confining stresses, soil structure (fabric, age and cementation), the magnitude and duration of the cyclic loading, and the extent to which shear stress reversal occurs.

Earthquake Liquefaction

The pressures generated during large earthquakes with many cycles of shaking can cause the liquefied sand and excess water to force its way to the ground surface from several meters below the ground. This is often observed as "sand boils" also called "sand blows" or "sand volcanoes" (as they appear to form small volcanic craters) at the ground surface. The phenomenon may incorporate both flow of already liquefied sand from a layer below ground, and a quicksand effect whereby upward flow of water initiates liquefaction in overlying non-liquefied sandy deposits due to buoyancy.

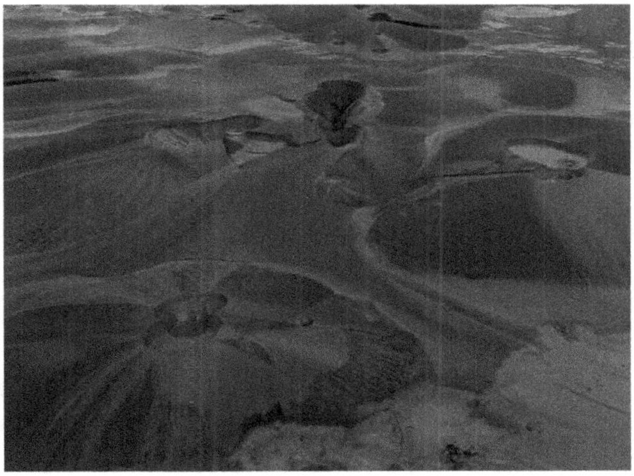

Sand boils that erupted during the 2011 Christchurch earthquake.

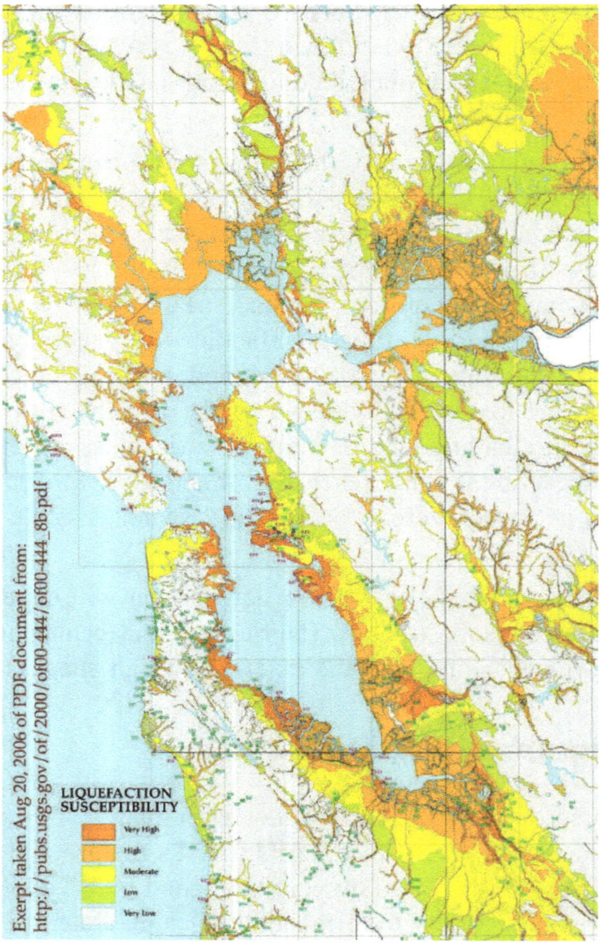

A liquefaction susceptibility map – excerpt of USGS map for the San Francisco Bay Area. Many areas of concern in this region are also densely urbanized.

The other common observation is land instability – cracking and movement of the ground down slope or towards unsupported margins of rivers, streams, or the coast. The failure of ground in this manner is called 'lateral spreading', and may occur on very shallow slopes of angles of only 1 or 2 degrees from the horizontal.

One positive aspect of soil liquefaction is the tendency for the effects of earthquake shaking to be significantly damped (reduced) for the remainder of the earthquake. This is because liquids do not support a shear stress and so once the soil liquefies due to shaking, subsequent earthquake shaking (transferred through ground by shear waves) is not transferred to buildings at the ground surface.

Studies of liquefaction features left by prehistoric earthquakes, called paleoliquefaction or paleoseismology, can reveal a great deal of information about earthquakes that occurred before records were kept or accurate measurements could be taken.

Soil liquefaction induced by earthquake shaking is also a major contributor to urban seismic risk.

Effects

The effects of soil liquefaction on the built environment can be extremely damaging. Buildings whose foundations bear directly on sand which liquefies will experience a sudden loss of support, which will result in drastic and irregular settlement of the building causing structural damage, including cracking of foundations and damage to the building structure itself, or may leave the structure unserviceable afterwards, even without structural damage. Where a thin crust of non-liquefied soil exists between building foundation and liquefied soil, a 'punching shear' type foundation failure may occur. The irregular settlement of ground may also break underground utility lines. The upward pressure applied by the movement of liquefied soil through the crust layer can crack weak foundation slabs and enter buildings through service ducts, and may allow water to damage the building contents and electrical services.

The effects of lateral spreading (River Road in Christchurch following the 2011 Christchurch earthquake)

Damage in Brooklands from the 2010 Canterbury earthquake, where buoyancy caused by soil liquefaction pushed up an underground service including this manhole

Bridges and large buildings constructed on pile foundations may lose support from the adjacent soil and buckle, or come to rest at a tilt after shaking.

Sloping ground and ground next to rivers and lakes may slide on a liquefied soil layer (termed 'lateral spreading'), opening large cracks or fissures in the ground, and can cause significant damage to buildings, bridges, roads and services such as water, natural gas, sewerage, power and telecommunications installed in the affected ground. Buried tanks and manholes may float in the liquefied soil due to buoyancy. Earth embankments such as flood levees and earth dams may lose stability or collapse if the material comprising the embankment or its foundation liquefies.

Over geolocic time, liquefaction of soil material due to earthquakes could provide a dense parent material in which the fragipan may develop through pedogenesis.

Mitigation Methods

Methods to mitigate the effects of soil liquefaction have been devised by earthquake engineers and include various soil compaction techniques such as vibro compaction (compaction of the soil by depth vibrators), dynamic compaction, and vibro stone columns. These methods result in the densification of soil and enable buildings to withstand soil liquefaction.

Existing buildings can be mitigated by injecting grout into the soil to stabilize the layer of soil that is subject to liquefaction.

Quicksand

Quicksand forms when water saturates an area of loose sand and the ordinary sand is agitated. When the water trapped in the batch of sand cannot escape, it creates liquefied soil that can no longer support weight. Quicksand can be formed by standing or (upwards) flowing underground water (as from an underground spring), or by earthquakes. In the case of flowing underground water, the force of the water flow opposes the force of gravity, causing the granules of sand to be more buoyant. In the case of earthquakes, the shaking force can increase the pressure of shallow groundwater, liquefying sand and silt deposits. In both cases, the liquefied surface loses strength, causing buildings or other objects on that surface to sink or fall over.

The saturated sediment may appear quite solid until a change in pressure or shock initiates the liquefaction, causing the sand to form a suspension with each grain surrounded by a thin film of water. This cushioning gives quicksand, and other liquefied sediments, a spongy, fluidlike texture. Objects in the liquefied sand sink to the level at which the weight of the object is equal to the weight of the displaced sand/water mix and the object *floats* due to its buoyancy.

Quick Clay

Quick clay, also known as *Leda Clay* in Canada, is a water-saturated gel, which in its solid form resemble a unique form of highly sensitive clay. This clay has a tendency to change from a relatively stiff condition to a liquid mass when it is disturbed. This gradual change in appearance from solid to liquid is a process known as spontaneous liquefaction. The clay retains a solid structure despite the high water content (up to 80% by volume), because surface tension holds water-coated flakes of clay together in a delicate structure. When the structure is broken by a shock or sufficient shear, it turns to a fluid state.

Quick clay is only found in the northern countries such as Russia, Canada, Alaska in the U.S., Norway, Sweden, and Finland, which were glaciated during the Pleistocene epoch.

Quick clay has been the underlying cause of many deadly landslides. In Canada alone, it has been associated with more than 250 mapped landslides. Some of these are ancient, and may have been triggered by earthquakes.

Turbidity Currents

Submarine landslides are turbidity currents and consist of water saturated sediments flowing downslope. An example occurred during the 1929 Grand Banks earthquake that struck the continental slope off the coast of Newfoundland. Minutes later, transatlantic telephone cables began breaking sequentially, farther and farther downslope, away from the epicenter. Twelve cables were snapped in a total of 28 places. Exact times and locations were recorded for each break. Investigators suggested that a 60-mile-per-hour (100 km/h) submarine *landslide* or turbidity current of water saturated sediments swept 400 miles (600 km) down the continental slope from the earthquake's epicenter, snapping the cables as it passed.

RESPONSE SPECTRUM

A response spectrum is simply a plot of the peak or steady-state response (displacement, velocity or acceleration) of a series of oscillators of varying natural frequency, that are forced into motion by the same base vibration or shock. The resulting plot can then be used to pick off the response of any linear system, given its natural frequency of oscillation. One such use is in assessing the peak response of buildings to earthquakes. The science of strong ground motion may use some values from the ground response spectrum (calculated from recordings of surface ground motion from seismographs) for correlation with seismic damage.

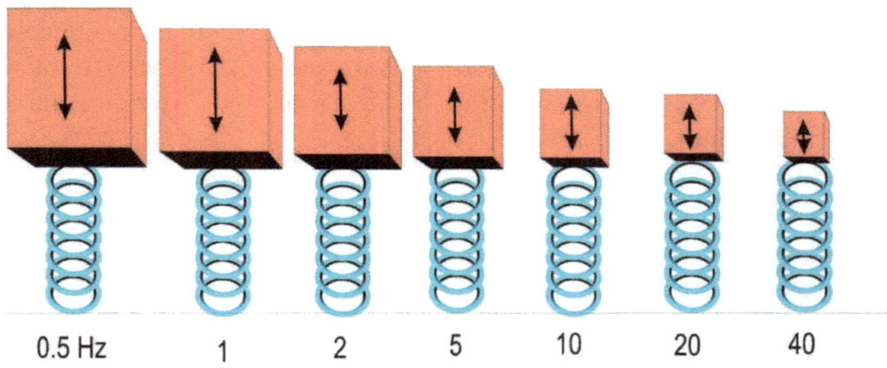

A series of mixed vertical oscillators

If the input used in calculating a response spectrum is steady-state periodic, then the steady-state result is recorded. Damping must be present, or else the response will be infinite. For transient input (such as seismic ground motion), the peak response is reported. Some level of damping is generally assumed, but a value will be obtained even with no damping.

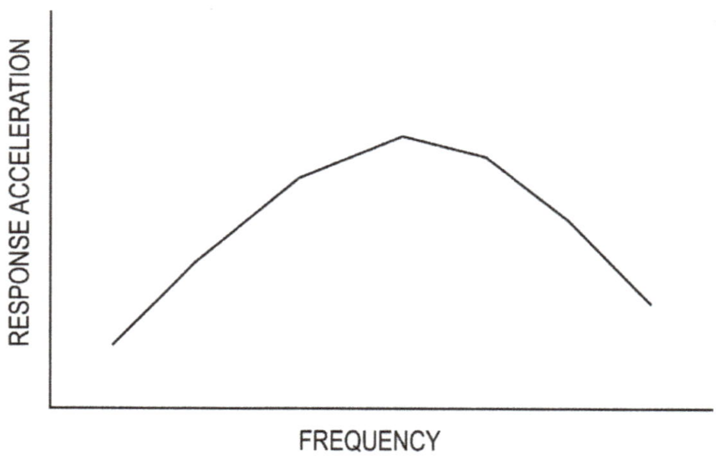

A plot of the peak acceleration for the mixed vertical oscillators

Response spectra can also be used in assessing the response of linear systems with multiple modes of oscillation (multi-degree of freedom systems), although they are only accurate for low levels of damping. Modal analysis is performed to identify the modes, and the response in that mode can be picked from the response spectrum. These peak responses are then combined to estimate a total response. A typical combination method is the square root of the sum of the squares (SRSS) if the modal frequencies are not close. The result is typically different from that which would be calculated directly from an input, since phase information is lost in the process of generating the response spectrum.

The main limitation of response spectra is that they are only universally applicable for linear systems. Response spectra can be generated for non-linear systems, but are only applicable to systems with the same non-linearity, although attempts have been made to develop non-linear seismic design spectra with wider structural application. The results of this cannot be directly combined for multi-mode response.

Seismic Response Spectra

Response spectra are very useful tools of earthquake engineering for analyzing the performance of structures and equipment in earthquakes, since many behave principally as simple oscillators (also known as single degree of freedom systems). Thus, if you can find out the natural frequency of the structure, then the peak response of the building can be estimated by reading the value from the ground response spectrum for the appropriate frequency. In most building codes in seismic regions, this value forms the basis for calculating the forces that a structure must be designed to resist (seismic analysis).

As mentioned earlier, the ground response spectrum is the response plot done at the free surface of the earth. Significant seismic damage may occur if the building response is 'in tune' with components of the ground motion (resonance), which may be identified from the response spectrum. This was observed in the 1985 Mexico City Earthquake where the oscillation of the deep-soil lake bed was similar to the natural frequency of mid-rise concrete buildings, causing significant damage. Shorter (stiffer) and taller (more flexible) buildings suffered less damage.

In 1941 at Caltech, George W. Housner began to publish calculations of response spectra from accelerographs. In the 1982 EERI Monograph on "Earthquake Design and Spectra", Newmark and Hall describe how they developed an "idealized" seismic response spectrum based on a range of response spectra generated for available earthquake records. This was then further developed into a design response spectrum for use in structural design, and this basic form (with some modifications) is now the basis for structural design in seismic regions throughout the world (typically plotted against structural "period", the inverse of frequency). A nominal level of damping is assumed (5% of critical damping).

For "regular" low-rise buildings, the structural response to earthquakes is characterized by the fundamental mode (a "waving" back-and-forth), and most building codes permit design forces to be calculated from the design spectrum on the basis of that frequency, but for more complex structures, combination of the results for many modes (calculated through modal analysis) is often required. In extreme cases, where structures are either too irregular, too tall or of significance to a community in disaster response, the response spectrum approach is no longer appropriate, and more complex analysis is required, such as non-linear static or dynamic analysis like in seismic performance analysis technique.

SEISMIC HAZARD

A seismic hazard is the probability that an earthquake will occur in a given geographic area, within a given window of time, and with ground motion intensity exceeding a given threshold. With a hazard thus estimated, risk can be assessed and included in such areas as building codes for standard buildings, designing larger buildings and infrastructure projects, land use planning and determining insurance rates. The seismic hazard studies also may generate two standard measures of anticipated ground motion, both confusingly abbreviated MCE; the simpler probabilistic Maximum Considered Earthquake (or Event), used in standard building codes, and the more detailed and deterministic Maximum Credible Earthquake incorporated in the design of larger buildings and civil infrastructure like dams or bridges. It is important to clarify which MCE is being discussed.

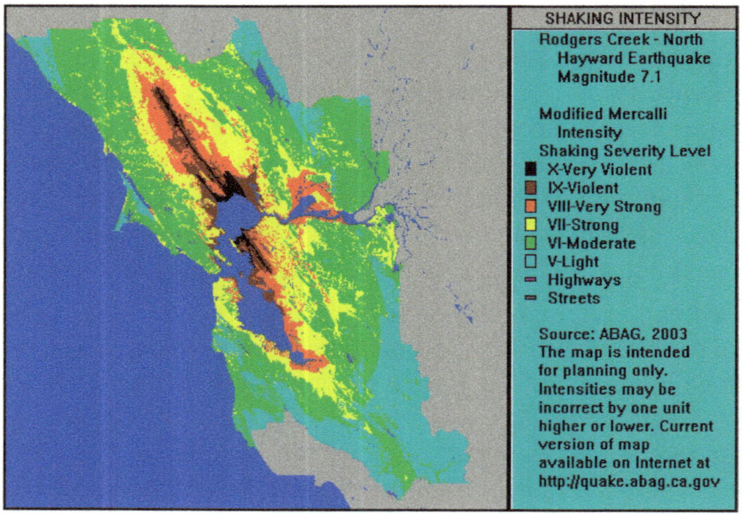

Surface motion map for a hypothetical earthquake on the northern portion of the Hayward Fault Zone and its presumed northern extension, the Rodgers Creek Fault Zone

Calculations for determining seismic hazard were first formulated by C. Allin Cornell in 1968 and, depending on their level of importance and use, can be quite complex. The regional geology and seismology setting is first examined for sources and patterns of earthquake occurrence, both in depth and at the surface from seismometer records; secondly, the impacts from these sources are assessed relative to local geologic rock and soil types, slope angle and groundwater conditions. Zones of similar potential earthquake shaking are thus determined and drawn on maps. The well known San Andreas Fault is illustrated as a long narrow elliptical zone of greater potential motion, like many areas along continental margins associated with the Pacific ring of fire. Zones of higher seismicity in the continental interior may be the site for intraplate earthquakes) and tend to be drawn as broad areas, based on historic records, like the 1812 New Madrid earthquake, since specific causative faults are generally not identified as earthquake sources.

Each zone is given properties associated with source potential: how many earthquakes per year, the maximum size of earthquakes (maximum magnitude), etc. Finally, the calculations require formulae that give the required hazard indicators for a given earthquake size and distance. For example, some districts prefer to use peak acceleration, others use peak velocity, and more sophisticated uses require response spectral ordinates.

The computer program then integrates over all the zones and produces probability curves for the key ground motion parameter. The final result gives a 'chance' of exceeding a given value over a specified amount of time. Standard building codes for homeowners might be concerned with a 1 in 500 years chance, while nuclear plants look at the 10,000 year time frame. A longer-term seismic history can be obtained through paleoseismology. The results may be in the form of a ground response spectrum for use in seismic analysis.

More elaborate variations on the theme also look at the soil conditions. Higher ground motions are likely to be experienced on a soft swamp compared to a hard rock site. The standard seismic hazard calculations become adjusted upwards when postulating characteristic earthquakes. Areas with high ground motion due to soil conditions are also often subject to soil failure due to liquefaction. Soil failure can also occur due to earthquake-induced landslides in steep terrain. Large area landsliding can also occur on rather gentle slopes as was seen in the Good Friday earthquake in Anchorage, Alaska, March 28, 1964.

MCES

In a normal seismic hazard analyses intended for the public, that of a "maximum considered earthquake", or "maximum considered event" (MCE) for a specific area, is an earthquake that is expected to occur once in approximately 2,500 years; that is, it has a 2-percent probability of being exceeded in 50 years. The term is used specifically for general building codes, which people commonly occupy; building codes in many localities will require non-essential buildings to be designed for "collapse prevention" in an MCE, so that the building remains standing - allowing for safety and escape of occupants - rather than full structural survival of the building.

A far more detailed and stringent MCE stands for "maximum credible earthquake", which is used in designing for skyscrapers and larger civil infrastructure, like dams, where structural failure could lead to other catastrophic consequences. These MCEs might require determining more than one specific earthquake event, depending on the variety of structures included.

US Seismic Hazard Maps

Some maps released by the USGS are shown with peak ground acceleration with a 10% probability of exceedance in 50 years, measured in Metre per second squared. For parts of the US, the National Seismic Hazard Mapping Project in 2008 resulted in seismic hazard maps showing peak acceleration (as a percentage of gravity) with a 2% probability of exceedance in 50 years.

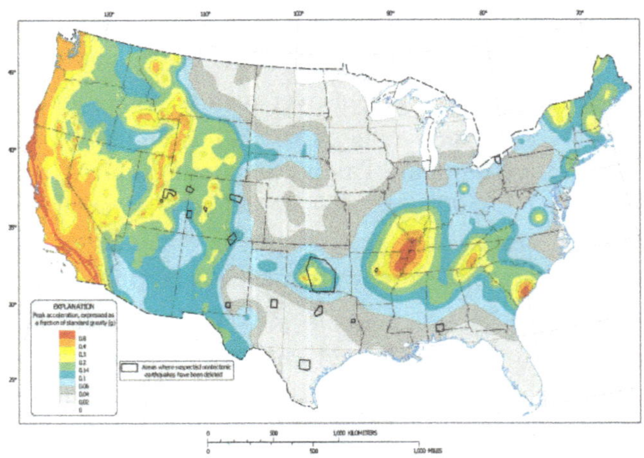

Two-percent probability of exceedance in 50 years map of peak ground acceleration

Map of peak ground acceleration with 2% probability of exceedance in 50 years

Temblor, a company founded in 2014, offers a seismic hazard rank for the all of the conterminous US. This service is free and ad-free for the public. The hazard rank "is made for the likelihood of experiencing strong shaking (0.4g peak ground acceleration) in 30 years, based on the 2014 USGS NSHMP hazard model."

GROUND–STRUCTURE INTERACTION

Design

The interaction between soil (ground) and structure consists of an exchange of mutual stress between the structure itself and the foundations ground. This means that the movement of the whole ground-structure system is influenced by the type of ground as well as by the type of structure. The main types of foundations, based upon several building characteristics, are:

- Isolated plinths (currently not feasible)
- Plinths connected by foundations beams
- Reverse beams
- A plate (used for low-quality grounds)

The filing of foundations grounds takes place according to the mechanical properties of the grounds themselves: in Italy, for instance, according to the new earthquake-proof norm – Ordinanza 3274/2003 – you can identify the following categories:

- Category A: homogeneous rock formations
- Category B: compact granular or clayey soil
- Category C: quite compact granular or clayey soil

- Category D: not much compact granular or clayey soil
- Category E: alluvial surface layer grounds (very low quality soil)

The type of foundations is selected according to the type of ground; for instance, in the case of homogeneous rock formations connected plinths are selected, while in the case of very low quality grounds plates are chosen.

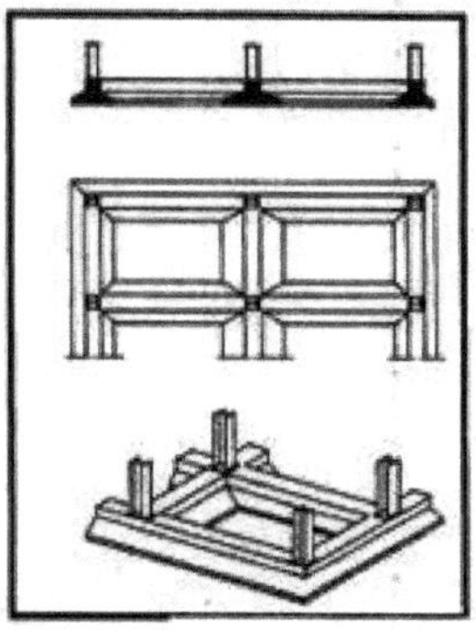

Reverse beams

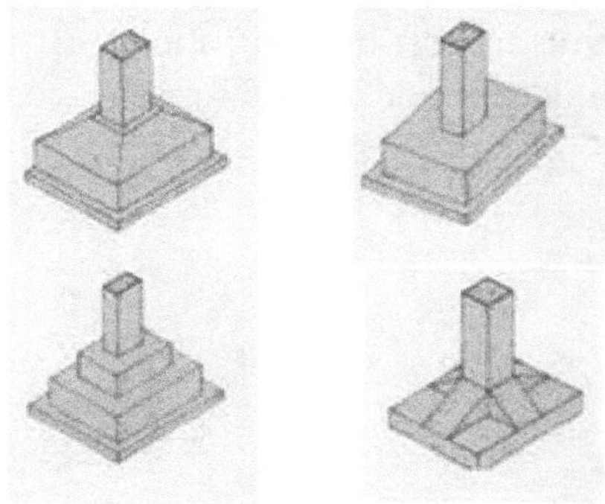

Isolated plinths

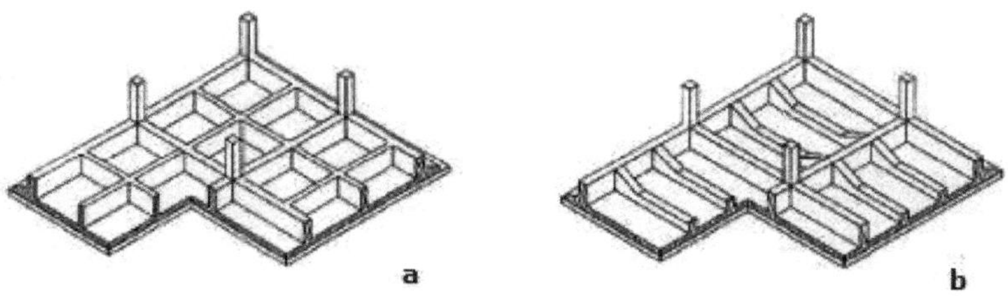

Plate

Both grounds and structures can be more or less deformable; their combination can or cannot cause the amplification of the seismic effects on the structure. Ground, in fact, is a filter with respect to all the main seismic waves, as stiffer soil fosters high-frequency seismic waves while less compact soil accommodates lower frequency waves. Therefore, a stiff building, characterized by a high fundamental frequency, suffers amplified damage when built on stiff ground and then subjected to higher frequencies.

For instance, suppose there are two buildings that share the same high stiffness. They stand on two different soil types: the first, stiff and rocky—the second, sandy and deformable. If subjected to the same seismic event, the building on the stiff ground suffers greater damage.

The second interaction effect, tied to mechanical properties of soil, is about the lowering (sinking) of foundations, worsened by the seismic event itself, especially about less compact grounds. This phenomenon is called soil liquefaction.

Mitigation

The methods most used to mitigate the problem of the ground-structure interaction consist of the employment of the before-seen isolation systems and of some ground brace techniques, which are adopted above all on the low-quality ones (categories D and E). The most diffused techniques are the jet-grouting technique and the pile work technique. The jet-grouting technique consists of injecting in the subsoil some liquid concrete by means of a drill. When this concrete hardens it forms a sort of column that consolidates the surrounding soil. This process is repeated on all areas of the structure. The pile work technique consists of using piles, which, once inserted in the ground, support the foundation and the building above, by moving the loads or the weights towards soil layers that are deeper and therefore more compact and movement-resistant.

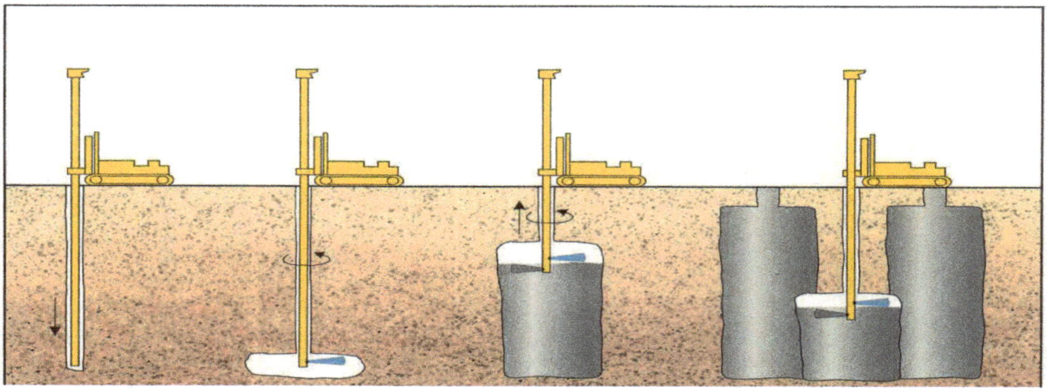

REFERENCES

1. Holtz, R. and Kovacs, W. (1981), An Introduction to Geotechnical Engineering, Prentice-Hall, Inc. ISBN 0-13-484394-0
2. Bowles, J. (1988), Foundation Analysis and Design, McGraw-Hill Publishing Company. ISBN 0-07-006776-7
3. Cedergren, Harry R. (1977), Seepage, Drainage, and Flow Nets, Wiley. ISBN 0-471-14179-8
4. Kramer, Steven L. (1996), Geotechnical Earthquake Engineering, Prentice-Hall, Inc. ISBN 0-13-374943-6
5. Freeze, R.A. & Cherry, J.A., (1979), Groundwater, Prentice-Hall. ISBN 0-13-365312-9
6. Lunne, T. & Long, M.,(2006), Review of long seabed samplers and criteria for new sampler design, Marine Geology, Vol 226, p. 145-165
7. Mitchell, James K. & Soga, K. (2005), Fundamentals of Soil Behavior 3rd ed., John Wiley & Sons, Inc. ISBN 978-0-471-46302-3
8. Rajapakse, Ruwan., (2005), "Pile Design and Construction", 2005. ISBN 0-9728657-1-3
9. Fang, H.-Y. and Daniels, J. (2005) Introductory Geotechnical Engineering : an environmental perspective, Taylor & Francis. ISBN 0-415-30402-4

10. NAVFAC (Naval Facilities Engineering Command) (1986) Design Manual 7.01, Soil Mechanics, US Government Printing Office
11. NAVFAC (Naval Facilities Engineering Command) (1986) Design Manual 7.02, Foundations and Earth Structures, US Government Printing Office
12. NAVFAC (Naval Facilities Engineering Command) (1983) Design Manual 7.03, Soil Dynamics, Deep Stabilization and Special Geotechnical Construction, US Government Printing Office
13. Terzaghi, K., Peck, R.B. and Mesri, G. (1996), Soil Mechanics in Engineering Practice 3rd Ed., John Wiley & Sons, Inc. ISBN 0-471-08658-4
14. Santamarina, J.C., Klein, K.A., & Fam, M.A. (2001), "Soils and Waves: Particulate Materials Behavior, Characterization and Process Monitoring", Wiley, ISBN 978-0-471-49058-6
15. Firuziaan, M. and Estorff, O., (2002), "Simulation of the Dynamic Behavior of Bedding-Foundation-Soil in the Time Domain", Springer Verlag.
16. Hazen, A. (1920). Transactions of the American Society of Civil Engineers. 83: 1717–1745. Missing or empty |title= (help)
17. "Geologists arrive to study liquefaction". One News. 10 September 2010. Retrieved 12 November 2011.
18. "Christchurch areas to be abandoned". The New Zealand Herald. NZPA. 7 March 2011. Retrieved 12 November 2011.
19. NEHRP recommended provisions for seismic regulations for new buildings and other structures (FEMA 450). Washington D.C.: National Institute of Building Sciences. 2004. |first1= missing |last1= in Authors list (help)
20. EN1998-5:2004 Eurocode 8 – Design of structures for earthquake resistance. Part 5: Foundations, retaining structures and geotechnical aspects. Brussels: European Committee for Standardisation. 2004. |first1= missing |last1= in Authors list (help)
21. International Code Council Inc. (ICC) (2006). International Building Code. Birmingham, Alabama: International Conference of Building Officials, and Southern Building Code Congress International, Inc. p. 679. ISBN 978-1-58001-302-4.
22. Casagrande, Arthur (1976). "Liquefaction and cyclic deformation of sands: A critical review". Harvard Soil Mechanics Series No. 88.
23. Jefferies, Mike; Been, Ken (2006). Soil Liquefaction: A Critical State Approach. Taylor & Francis. ISBN 978-0-419-16170-7.
24. Youd, T. L.; Member, Asce, I. M. Idriss, Chair; Fellow, Asce, Ronald D. Andrus, Co-Chair; Arango, Ignacio; Castro, Gonzalo; Christian, John T.; Dobry, Richardo; Finn, W. D. Liam; et al. (2001). "Liquefaction Resistance of Soils: Summary Report from the 1996 NCEER and 1998 NCEER/NSF Workshops on Evaluation of Liquefaction Resistance of Soils". Journal of Geotechnical and Geoenvironmental Engineering. 127 (10): 297–313. doi:10.1061/(ASCE)1090-0241(2001)127:10(817).

25. Robertson, P.K., and Fear, C.E. (1995). "Liquefaction of sands and its evaluation.", Proceedings of the 1st International Conference on Earthquake Geotechnical Engineering, Tokyo
26. Robertson, P K; Wride, CE (Fear) (1998). "Evaluating cyclic liquefaction potential using the cone penetration test". Canadian Geotechnical Journal. 35 (3): 442–59. doi:10.1139/t98-017.
27. http://earthquake.usgs.gov/research/hazmaps/whats_new/workshops/CEUS-WORKSHP/Tuesday/NE-Tuttle.2.pdf
28. "Liquefaction", Institution of Professional Engineers of New Zealand
29. Scalenghe, R., Certini, G., Corti G., Zanini E., Ugolini, F.C. (2004). "Segregated ice and liquefaction effects on compaction of fragipans". Soil Science Society of America Journal. 68: 204–214. doi:10.2136/sssaj2004.2040.
30. Liquefaction Mitigation
31. R. Lukas and B. Moore, Dynamic Compaction
32. "Geoscape Ottawa-Gatineau Landslides", Natural Resources Canada
33. Heezen, B. C.; Ewing, W. M. (1952). "Turbidity currents and submarine slumps, and the 1929 Grand Banks [Newfoundland] earthquake". American Journal of Science. 250 (12): 849–73. doi:10.2475/ajs.250.12.849.

CHAPTER 10

Seismic Retrofitting

Chapter Contents

A Seismic Retrofit provides existing structures with more resistance to seismic activity due to earthquakes. In buildings, this process typically includes strengthening weak connections found in roof to wall connections, continuity ties, shear walls and the roof diaphragm.

INTRODUCTION

Seismic retrofitting is the modification of existing structures to make them more resistant to seismic activity, ground motion, or soil failure due to earthquakes. With better understanding of seismic demand on structures and with our recent experiences with large earthquakes near urban centers, the need of seismic retrofitting is well acknowledged. Prior to the introduction of modern seismic codes in the late 1960s for developed countries (US, Japan etc.) and late 1970s for many other parts of the world (Turkey, China etc.), many structures were designed without adequate detailing and reinforcement for seismic protection. In view of the imminent problem, various research work has been carried out. State-of-the-art technical guidelines for seismic assessment, retrofit and rehabilitation have been published around the world - such as the ASCE-SEI 41and the New Zealand Society for Earthquake

Engineering (NZSEE)'s guidelines. These codes must be regularly updated; the 1994 Northridge earthquake brought to light the brittleness of welded steel frames, for example.

Infill shear trusses — University of California dormitory, Berkeley

External bracing of an existing reinforced concrete parking garage (Berkeley)

Port Authority Bus Terminal

The retrofit techniques outlined here are also applicable for other natural hazards such as tropical cyclones, tornadoes, and severe winds from thunderstorms. Whilst current practice of seismic retrofitting is predominantly concerned with structural improvements to reduce the seismic hazard of using the structures, it is similarly essential to reduce the hazards and losses from non-structural elements. It is also important to keep in mind that there is no such thing as an earthquake-proof structure, although seismic performance can be greatly enhanced through proper initial design or subsequent modifications.

STRATEGIES

Seismic retrofit (or rehabilitation) strategies have been developed in the past few decades following the introduction of new seismic provisions and the availability of advanced materials (e.g. fiber-reinforced polymers (FRP), fiber reinforced concrete and high strength steel). Retrofit strategies are different from retrofit techniques, where the former is the basic approach to achieve an overall retrofit performance objective, such as increasing strength, increasing deformability, reducing deformation demands while the latter is the technical methods to achieve that strategy, for example FRP jacketing.

- Increasing the global capacity (strengthening). This is typically done by the addition of cross braces or new structural walls.
- Reduction of the seismic demand by means of supplementary damping and/or use of base isolation systems.
- Increasing the local capacity of structural elements. This strategy recognizes the inherent capacity within the existing structures, and therefore adopt a more cost-effective approach to selectively upgrade local capacity (deformation/ductility, strength or stiffness) of individual structural components.
- Selective weakening retrofit. This is a counter intuitive strategy to change the inelastic mechanism of the structure, while recognizing the inherent capacity of the structure.

- Allowing sliding connections such as passageway bridges to accommodate additional movement between seismically independent structures.
- Addition of seismic friction dampers to brace weak structures and provide damping.

PERFORMANCE OBJECTIVES

In the past, seismic retrofit was primarily applied to achieve public safety, with engineering solutions limited by economic and political considerations. However, with the development of Performance based earthquake engineering (PBEE), several levels of performance objectives are gradually recognized:

- Public safety only. The goal is to protect human life, ensuring that the structure will not collapse upon its occupants or passersby, and that the structure can be safely exited. Under severe seismic conditions the structure may be a total economic write-off, requiring tear-down and replacement.
- Structure survivability. The goal is that the structure, while remaining safe for exit, may require extensive repair (but not replacement) before it is generally useful or considered safe for occupation. This is typically the lowest level of retrofit applied to bridges.
- Structure functionality. Primary structure undamaged and the structure is undiminished in utility for its primary application. A high level of retrofit, this ensures that any required repairs are only "cosmetic" - for example, minor cracks in plaster, drywall and stucco. This is the minimum acceptable level of retrofit for hospitals.
- Structure unaffected. This level of retrofit is preferred for historic structures of high cultural significance.

TECHNIQUES

Common seismic retrofitting techniques fall into several categories:

One of many "earthquake bolts" found throughout period houses in the city of Charleston subsequent to the Charleston earthquake of 1886. They could be tightened and loosened to support the house without having to otherwise demolish the house due to instability. The bolts were directly loosely connected to the supporting frame of the house.

External Post-Tensioning

The use of external post-tensioning for new structural systems have been developed in the past decade. Under the PRESS (Precast Seismic Structural Systems), a large-scale U.S./Japan joint research program, unbonded post-tensioning high strength steel tendons have been used to achieve a moment-resisting system that has self-centering capacity. An extension of the same idea for seismic retrofitting has been experimentally tested for seismic retrofit of California bridges under a Caltrans research project and for seismic retrofit of non-ductile reinforced concrete frames. Pre-stressing can increase the capacity of structural elements such as beam, column and beam-column joints. It should be noted that external pre-stressing has been used for structural upgrade for gravity/live loading since the 1970s.

Base Isolators

Base isolation is a collection of structural elements of a building that should substantially decouple the building's structure from the shaking ground thus protecting the building's integrity and enhancing its seismic performance. This earthquake engineering technology, which is a kind of seismic vibration control, can be applied both to a newly designed building and to seismic upgrading of existing structures. Normally, excavations are made around the building and the building is separated from the foundations. Steel or reinforced concrete beams replace the connections to the foundations, while under these, the isolating pads, or base isolators, replace the material removed. While the base isolation tends to restrict transmission of the ground motion to the building, it also keeps the building positioned properly over the foundation. Careful attention to detail is required where the building interfaces with the ground, especially at entrances, stairways and ramps, to ensure sufficient relative motion of those structural elements.

Supplementary Dampers

Supplementary dampers absorb the energy of motion and convert it to heat, thus "damping" resonant effects in structures that are rigidly attached to the ground. In addition to adding energy dissipation capacity to the structure, supplementary damping can reduce the displacement and acceleration demand within the structures. In some cases, the threat of damage does not come from the initial shock itself, but rather from the periodic resonant motion of the structure that repeated ground motion induces. In the practical sense, supplementary dampers act similarly to Shock absorbers used in automotive suspensions.

Tuned Mass Dampers

Tuned mass dampers (TMD) employ movable weights on some sort of springs. These are typically employed to reduce wind sway in very tall, light buildings. Similar designs may be employed to impart earthquake resistance in eight to ten story buildings that are prone to destructive earthquake induced resonances.

Slosh Tank

A slosh tank is a large tank of fluid placed on an upper floor. During a seismic event, the fluid in this tank will slosh back and forth, but is directed by baffles - partitions that prevent the tank itself becoming resonant; through its mass the water may change or counter the resonant period of the building. Additional kinetic energy can be converted to heat by the baffles and is dissipated through the water - any temperature rise will be insignificant. One Rincon Hill in San Francisco is a recent skyscraper with a rooftop slosh tank.

Active Control System

Very tall buildings ("skyscrapers"), when built using modern lightweight materials, might sway uncomfortably (but not dangerously) in certain wind conditions. A solution to this problem is to include at some upper story a large mass, constrained, but free to move within a limited range, and moving on some sort of bearing system such as an air cushion or hydraulic film. Hydraulic pistons, powered by electric pumps and accumulators, are actively driven to counter the wind forces and natural resonances. These may also, if properly designed, be effective in controlling excessive motion - with or without applied power - in an earthquake. In general, though, modern steel frame high rise buildings are not as subject to dangerous motion as are medium rise (eight to ten story) buildings, as the resonant period of a tall and massive building is longer than the approximately one second shocks applied by an earthquake.

Adhoc Addition of Structural Support/Reinforcement

The most common form of seismic retrofit to lower buildings is adding strength to the existing structure to resist seismic forces. The strengthening may be limited to connections between existing building elements or it may involve adding primary resisting elements such as walls or frames, particularly in the lower stories.

Connections between Buildings and their Expansion Additions

Frequently, building additions will not be strongly connected to the existing structure, but simply placed adjacent to it, with only minor continuity in flooring, siding, and roofing. As a result, the addition may have a different resonant period than the original structure, and they may easily detach from one another. The relative motion will then cause the two parts to collide, causing severe structural damage. Seismic modification will either tie the two building components rigidly together so that they behave as a single mass or it will employ dampers to expend the energy from relative motion, with appropriate allowance for this motion, such as increased spacing and sliding bridges between sections.

Exterior Reinforcement of Building

Exterior Concrete Columns

Historic buildings, made of unreinforced masonry, may have culturally important interior detailing or murals that should not be disturbed. In this case, the solution may be to add a number of steel, reinforced concrete, or poststressed concrete columns to the exterior. Careful attention must be paid to the connections with other members such as footings, top plates, and roof trusses.

Infill Shear Trusses

Shown here is an exterior shear reinforcement of a conventional reinforced concrete dormitory building. In this case, there was sufficient vertical strength in the building columns and sufficient shear strength in the lower stories that only limited shear reinforcement was required to make it earthquake resistant for this location near the Hayward fault.

Massive Exterior Structure

In other circumstances, far greater reinforcement is required. In the structure shown at right — a parking garage over shops — the placement, detailing, and painting of the reinforcement becomes itself an architectural embellishment.

TYPICAL RETROFIT SCENARIO AND SOLUTION

Soft-Story Failure

This collapse mode is known as *soft story collapse*. In many buildings the ground level is designed for different uses than the upper levels. Low rise residential structures may be built over a parking garage which have large doors on one side. Hotels may have a tall ground floors to allow for a grand entrance or ballrooms. Office buildings may have stores in the ground floor which desire continuous windows for display.

Partial failure due to inadequate shear structure at garage level. Damage in San Francisco due to the Loma Prieta event.

Traditional seismic design assumes that the lower stories of a building are stronger than the upper stories and where this is not the case—if the lower story is less strong than the upper structure—the structure will not respond to earthquakes in the expected fashion. Using modern design methods, it is possible to take a weak story into account. Several failures of this type in one large apartment complex caused most of the fatalities in the 1994 Northridge earthquake.

Typically, where this type of problem is found, the weak story is reinforced to make it stronger than the floors above by adding shear walls or moment frames. Moment frames consisting of inverted **U** bents are useful in preserving lower story garage access, while a lower cost solution may be to use shear walls or trusses in several locations, which partially reduce the usefulness for automobile parking but still allow the space to be used for other storage.

Beam-Column Joint Connections

Beam-column joint connections are a common structural weakness in dealing with seismic retrofitting. Prior to the introduction of modern seismic codes in early 1970s, beam-column joints were typically non-engineered or designed. Laboratory testings have confirmed the seismic vulnerability of these poorly detailed and under-designed connections. Failure of beam-column joint connections can typically lead to catastrophic collapse of a frame-building, as often observed in recent earthquakes

Corner joint steel reinforcement and high tensile strength rods with grouted anti-burst jacket below

For reinforced concrete beam-column joints - various retrofit solutions have been proposed and tested in the past 20 years. Philosophically, the various seismic retrofit strategies discussed above can be implemented for reinforced concrete joints. Concrete or steel jacketing have been a popular retrofit technique until the advent of composite materials such as Carbon fiber-reinforced polymer (FRP). Composite materials such as carbon FRP and aramic FRP have been extensively tested for use in seismic retrofit with some success. One novel technique includes the use of selective weakening of the beam and added external post-tensioning to the joint in order to achieve flexural hinging in the beam, which is more desirable in terms of seismic design.

Widespread weld failures at beam-column joints of low-to-medium rise steel buildings during the Northridge 1994 earthquake for example, have shown the structural defiencies of these 'modern-designed' post-1970s welded moment-resisting connections. A subsequent SAC research project has documented, tested and proposed several retrofit solutions for these welded steel moment-resisting connections. Various retrofit solutions have been developed for these welded joints - such as a) weld strengthening and b) addition of steel haunch or 'dog-bone' shape flange.

Following the Northridge earthquake, a number of steel moment -frame buildings were found to have experienced brittle fractures of beam to column connections. Discovery of these unanticipated brittle fractures of framing connections was alarming to engineers and the building industry. Starting in the 1960s, engineers began to regard welded steel moment-frame buildings as being among the most ductile systems contained in the building code. Many engineers believed that steel moment-frame buildings were essentially invulnerable to earthquake induced damage and thought that should damage occur, it would be limited to ductile yielding of members and connections. Observation of damage sustained by buildings in the 1994 Northridge earthquake indicated that contrary to the intended behavior, in many cases, brittle fractures initiated within the connections at very low levels of plastic demand. In September, 1994, The SAC joint Venture, AISC, AISI, and NIST jointly convened an international workshop in Los Angeles to coordinate the efforts of various participants and to lay the foundation for systematic investigation and resolution of the problem. In September 1995 the SAC Joint Venture entered into a contractual agreement with FEMA to conduct Phase II of the SAC Steel project. Under Phase II, SAC continued its extensive problem-focused study of the performance of moment resisting steel frames and connections of various configurations, with the ultimate goal of developing seismic design criteria for steel construction. As a result of these studies it is now known that the typical moment-resisting connection detail employed in steel moment frame construction prior to the 1994 Northridge earthquake had a number of features that rendered it inherently susceptible to brittle fracture.

Shear Failure within Floor Diaphragm

Floors in wooden buildings are usually constructed upon relatively deep spans of wood, called joists, covered with a diagonal wood planking or plywood to form a subfloor upon which the finish floor surface is laid. In many structures these are all aligned in the same direction. To prevent the beams from tipping over onto their side, blocking is used at each end, and for additional stiffness, blocking or diagonal wood or metal bracing may be placed between beams at one or more points in their spans. At the outer edge it is typical to use a single depth of blocking and a perimeter beam overall.

If the blocking or nailing is inadequate, each beam can be laid flat by the shear forces applied to the building. In this position they lack most of their original strength and the structure may further collapse. As part of a retrofit the blocking may be doubled, especially at the outer edges of the building. It may be appropriate to add additional nails between the sill plate of the perimeter wall erected upon the floor diaphragm, although this will require exposing the sill plate by removing interior plaster or exterior siding. As the sill plate may be quite old and dry and substantial nails must be used, it may be necessary to pre-drill a hole for the nail in the old wood to avoid splitting. When the wall is opened for this purpose it may also be appropriate to tie vertical wall elements into the foundation using specialty connectors and bolts glued with epoxy cement into holes drilled in the foundation.

Sliding Off Foundation and "Cripple Wall" Failure

Single or two story wood-frame domestic structures built on a perimeter or slab foundation are relatively safe in an earthquake, but in many structures built before 1950 the sill plate that sits between the concrete foundation and the floor diaphragm (perimeter foundation) or studwall (slab foundation) may not be sufficiently bolted in. Additionally, older attachments (without substantial corrosion-proofing) may have corroded to a point of weakness. A sideways shock can slide the building entirely off of the foundations or slab.

House slid off of foundation

Low cripple wall collapse and detachment of structure from concrete stairway

Often such buildings, especially if constructed on a moderate slope, are erected on a platform connected to a perimeter foundation through low stud-walls called "cripple wall" or *pin-up*. This low wall structure itself may fail in shear or in its connections to itself at the corners, leading to the building moving diagonally and collapsing the low walls. The likelihood of failure of the pin-up can be reduced by ensuring that the corners are well reinforced in shear and that the shear panels are well connected to each other through the corner posts. This requires structural grade sheet plywood, often treated for rot resistance. This grade of plywood is made without interior unfilled knots and with more, thinner layers than common plywood. New buildings designed to resist earthquakes will typically use OSB (oriented strand board), sometimes with metal joins between panels, and with well attached stucco covering to enhance its performance. In many modern tract homes, especially those built upon expansive (clay) soil the building is constructed upon a single and relatively thick monolithic slab, kept in one piece by high tensile rods that are stressed after the slab has set. This poststressing places the concrete under compression - a condition under which it is extremely strong in bending and so will not crack under adverse soil conditions.

Multiple Piers in Shallow Pits

Some older low-cost structures are elevated on tapered concrete pylons set into shallow pits, a method frequently used to attach outdoor decks to existing buildings. This is seen in conditions of damp soil, especially in tropical conditions, as it leaves a dry ventilated space under the house, and in far northern conditions of permafrost (frozen mud) as it keeps the building's warmth from destabilizing the ground beneath. During an earthquake, the pylons may tip, spilling the building to the ground. This can be overcome by using deep-bored holes to contain cast-in-place reinforced pylons, which are then secured to the floor panel at the corners of the building. Another technique is to add sufficient diagonal bracing or sections of concrete shear wall between pylons.

Reinforced Concrete Column Burst

Reinforced concrete columns typically contain large diameter vertical rebar (reinforcing bars) arranged in a ring, surrounded by lighter-gauge hoops of rebar. Upon analysis of failures due to earthquakes, it has been realized that the weakness was not in the vertical bars, but rather in inadequate strength and quantity of hoops. Once the integrity of the hoops is breached, the vertical rebar can flex outward, stressing the central column of concrete. The concrete then simply crumbles into small pieces, now unconstrained by the surrounding rebar. In new construction a greater amount of hoop-like structures are used.

Jacketed and grouted column on left, unmodified on right

One simple retrofit is to surround the column with a jacket of steel plates formed and welded into a single cylinder. The space between the jacket and the column is then filled with concrete, a process called grouting. Where soil or structure conditions require such additional modification, additional pilings may be driven near the column base and concrete pads linking the pilings to the pylon are fabricated at or below ground level. In the example shown not all columns needed to be modified to gain sufficient seismic resistance for the conditions expected. (This location is about a mile from the Hayward Fault Zone.)

Reinforced Concrete Wall Burst

Concrete walls are often used at the transition between elevated road fill and overpass structures. The wall is used both to retain the soil and so enable the use of a shorter span and also to transfer the weight of the span directly downward to footings in undisturbed soil. If these walls are inadequate they may crumble under the stress of an earthquake's induced ground motion.

One form of retrofit is to drill numerous holes into the surface of the wall, and secure short **L**-shaped sections of rebar to the surface of each hole with epoxy adhesive. Additional vertical and horizontal rebar is then secured to the new elements, a form is erected, and an additional layer of concrete is poured. This modification may be combined with additional footings in excavated trenches and additional support ledgers and tie-backs to retain the span on the bounding walls.

Brick Wall Resin and Glass Fiber Reinforcement

Brick building structures have been reinforced with coatings of glass fiber and appropriate resin (epoxy or polyester). In lower floors these may be applied over entire exposed surfaces, while in upper floors this may be confined to narrow areas around window and door openings. This application provides tensile strength that stiffens the wall against bending away from the side with the application. The efficient protection of an entire building requires extensive analysis and engineering to determine the appropriate locations to be treated.

Lift

Where moist or poorly consolidated alluvial soil interfaces in a "beach like" structure against underlying firm material, seismic waves traveling through the alluvium can be amplified, just as are water waves against a sloping beach. In these special conditions, vertical accelerations up to twice the force of gravity have been measured. If a building is not secured to a well-embedded foundation it is possible for the building to be thrust from (or with) its foundations into the air, usually with severe damage upon landing. Even if it is well-founded, higher portions such as upper stories or roof structures or attached structures such as canopies and porches may become detached from the primary structure.

Good practices in modern, earthquake-resistant structures dictate that there be good vertical connections throughout every component of the building, from undisturbed or engineered earth to foundation to sill plate to vertical studs to plate cap through each floor and continuing to the roof structure. Above the foundation and sill plate the connections are typically made using steel strap or sheet stampings, nailed to wood members using special hardened high-shear strength nails, and heavy angle stampings secured with through bolts, using large washers to prevent pull-through. Where inadequate bolts are provided between the sill plates and a foundation in existing construction (or are not trusted due to possible corrosion), special clamp plates may be added, each of which is secured to the foundation using expansion bolts inserted into holes drilled in an exposed face of concrete. Other members must then be secured to the sill plates with additional fittings.

Soil

One of the most difficult retrofits is that required to prevent damage due to soil failure. Soil failure can occur on a slope, a slope failure or landslide, or in a flat area due to liquefaction of water-saturated sand and/or mud. Generally, deep pilings must be driven into stable soil (typically hard mud or sand) or to underlying bedrock or the slope must be stabilized. For buildings built atop previous landslides the practicality of retrofit may be limited by economic factors, as it is not practical to stabilize a large, deep landslide. The likelihood of landslide or soil failure may also depend upon seasonal factors, as the soil may be more stable at the beginning of a wet season than at the beginning of the dry season. Such a "two season" *Mediterranean climate* is seen throughout California.

In some cases, the best that can be done is to reduce the entrance of water runoff from higher, stable elevations by capturing and bypassing through channels or pipes, and to drain water infiltrated directly and from subsurface springs by inserting horizontal perforated tubes. There are numerous locations in California where extensive developments have been built atop archaic landslides, which have not moved in historic times but which (if both water-saturated and shaken by an earthquake) have a high probability of moving *en masse*, carrying entire sections of suburban development to new locations. While the most modern of house structures (well tied to monolithic concrete foundation slabs reinforced with post tensioning cables) may survive such movement largely intact, the building will no longer be in its proper location.

Utility Pipes and Cables: Risks

Natural gas and propane supply pipes to structures often prove especially dangerous during and after earthquakes. Should a building move from its foundation or fall due to cripple wall collapse, the ductile iron pipes transporting the gas within the structure may be broken, typically at the location of threaded joints. The gas may then still be provided to the pressure regulator from higher pressure lines and so continue to flow in substantial quantities; it may then be ignited by a nearby source such as a lit pilot light or arcing electrical connection.

There are two primary methods of automatically restraining the flow of gas after an earthquake, installed on the low pressure side of the regulator, and usually downstream of the gas meter.

- A caged metal ball may be arranged at the edge of an orifice. Upon seismic shock, the ball will roll into the orifice, sealing it to prevent gas flow. The ball may later be reset by the use of an external magnet. This device will respond only to ground motion.

- A flow-sensitive device may be used to close a valve if the flow of gas exceeds a set threshold (very much like an electrical circuit breaker). This device will operate independently of seismic motion, but will not respond to minor leaks which may be caused by an earthquake.

It appears that the most secure configuration would be to use one of each of these devices in series.

Tunnels

Unless the tunnel penetrates a fault likely to slip, the greatest danger to tunnels is a landslide blocking an entrance. Additional protection around the entrance may be applied to divert any falling material (similar as is done to divert snow avalanches) or the slope above the tunnel may be stabilized in some way. Where only small- to medium-sized rocks and boulders are expected to fall, the entire slope may be covered with wire mesh, pinned down to the slope with metal rods. This is also a common modification to highway cuts where appropriate conditions exist.

Underwater Tubes

The safety of underwater tubes is highly dependent upon the soil conditions through which the tunnel was constructed, the materials and reinforcements used, and the maximum predicted earthquake expected, and other factors, some of which may remain unknown under current knowledge.

BART tube

A tube of particular structural, seismic, economic, and political interest is the BART (Bay Area Rapid Transit) transbay tube. This tube was constructed at the bottom of San Francisco Bay through an innovative process. Rather than pushing a shield through the soft bay mud, the tube was constructed on land in sections. Each section consisted of two inner train tunnels of circular cross section, a central access tunnel of rectangular cross section, and an outer oval shell encompassing the three inner tubes. The intervening space was filled with concrete. At the bottom of the bay a trench was excavated and a flat bed of crushed stone prepared to receive the tube sections. The sections were then floated into place and sunk, then joined with bolted connections to previously-placed sections. An overfill was then placed atop the tube to hold it down. Once completed from San Francisco to Oakland, the tracks and electrical components were installed. The predicted response of the tube during a major earthquake was likened to be as that of a string of (cooked) spaghetti in a bowl of gelatin dessert. To avoid overstressing the tube due to differential movements at each end, a sliding slip joint was included at the San Francisco terminus under the landmark Ferry Building.

The engineers of the construction consortium PBTB (Parsons Brinckerhoff-Tudor-Bechtel) used the best estimates of ground motion available at the time, now known to be insufficient given modern computational analysis methods and geotechnical knowledge. Unexpected settlement of the tube has reduced the amount of slip that can be accommodated without failure. These factors have resulted in the slip joint being designed too short to ensure survival of the tube under possible (perhaps even likely) large earthquakes in the region. To correct this deficiency the slip joint must be extended to allow for additional movement, a modification expected to be both expensive and technically and logistically difficult. Other retrofits to the BART tube include vibratory consolidation of the tube's overfill to avoid potential liquefying of the overfill, which has now been completed. (Should the overfill fail there is a danger of portions of the tube rising from the bottom, an event which could potentially cause failure of the section connections.)

BRIDGE RETROFIT

Bridges have several failure modes.

Expansion Rockers

Many short bridge spans are statically anchored at one end and attached to rockers at the other. This rocker gives vertical and transverse support while allowing the bridge span to expand and contract with temperature changes. The change in the length of the span is accommodated over a gap in the roadway by comb-like expansion joints. During severe ground motion, the rockers may jump from their tracks or be moved beyond their design limits, causing the bridge to unship from its resting point and then either become misaligned or fail completely. Motion can be constrained by adding ductile or high-strength steel restraints that are friction-clamped to beams and designed to slide under extreme stress while still limiting the motion relative to the anchorage.

Deck Rigidity

Suspension bridges may respond to earthquakes with a side-to-side motion exceeding that which was designed for wind gust response.

Additional diagonals were inserted under both decks of this bridge

Such motion can cause fragmentation of the road surface, damage to bearings, and plastic deformation or breakage of components. Devices such as hydraulic dampers or clamped sliding connections and additional diagonal reenforcement may be added.

Lattice Girders, Beams, and Ties

Lattice girders consist of two "I"-beams connected with a criss-cross lattice of flat strap or angle stock. These can be greatly strengthened by replacing the open lattice with plate members. This is usually done in concert with the replacement of hot rivets with bolts.

Obsolete riveted lattice members

Bolted plate lattice replacement, forming box members

Hot Rivets

Many older structures were fabricated by inserting red-hot rivets into pre-drilled holes; the soft rivets are then peened using an air hammer on one side and a bucking bar on the head end. As these cool slowly, they are left in an annealed (soft) condition, while the plate, having been hot rolled and quenched during manufacture, remains relatively hard. Under extreme stress the hard plates can shear the soft rivets, resulting in failure of the joint.

The solution is to burn out each rivet with an oxygen torch. The hole is then prepared to a precise diameter with a reamer. A special *locator bolt*, consisting of a head, a shaft matching the reamed hole, and a threaded end is inserted and retained with a nut, then tightened with a wrench. As the bolt has been formed from an appropriate high-strength alloy and has also been heat-treated, it is not subject to either the plastic shear failure typical of hot rivets nor the brittle fracture of ordinary bolts. Any partial failure will be in the plastic flow of the metal secured by the bolt; with proper engineering any such failure should be non-catastrophic.

Fill and Overpass

Elevated roadways are typically built on sections of elevated earth fill connected with bridge-like segments, often supported with vertical columns. If the soil fails where a bridge terminates, the bridge may become disconnected from the rest of the roadway and break away. The retrofit for this is to add additional reinforcement to any supporting wall, or to add deep caissons adjacent to the edge at each end and connect them with a supporting beam under the bridge.

Another failure occurs when the fill at each end moves (through resonant effects) in bulk, in opposite directions. If there is an insufficient founding shelf for the overpass, then it may fall. Additional shelf and ductile stays may be added to attach the overpass to the footings at one or both ends. The stays, rather than being fixed to the beams, may instead be clamped to them. Under moderate loading, these keep the overpass centered in the gap so that it is less likely to slide off its founding shelf at one end. The ability for the fixed ends to slide, rather than break, will prevent the complete drop of the structure if it should fail to remain on the footings.

Viaducts

Large sections of roadway may consist entirely of viaduct, sections with no connection to the earth other than through vertical columns. When concrete columns are used, the detailing is critical. Typical failure may be in the toppling of a row of columns due either to soil connection failure or to insufficient cylindrical wrapping with rebar. Both failures were seen in the 1995 Great Hanshin earthquake in Kobe, Japan, where an entire viaduct, centrally supported by a single row of large columns, was laid down to one side. Such columns are reinforced by excavating to the foundation pad, driving additional pilings, and adding a new, larger pad, well connected with rebar alongside or into the column. A column with insufficient wrapping bar, which is prone to burst and then hinge at the bursting point, may be completely encased in a circular or elliptical jacket of welded steel sheet and grouted as described above.

Cypress Freeway viaduct collapse. Note failure of inadequate anti-burst wrapping and lack of connection between upper and lower vertical elements.

Sometimes viaducts may fail in the connections between components. This was seen in the failure of the Cypress Freeway in Oakland, California, during the Loma Prieta earthquake. This viaduct was a two-level structure, and the upper portions of the columns were not well connected to the lower portions that supported the lower level; this caused the upper deck to collapse upon the lower deck. Weak connections such as these require additional external jacketing - either through external steel components or by a complete jacket of reinforced concrete, often using stub connections that are glued (using epoxy adhesive) into numerous drilled holes. These stubs are then connected to additional wrappings, external forms (which may be temporary or permanent) are erected, and additional concrete is poured into the space. Large connected structures similar to the Cypress Viaduct must also be properly analyzed in their entirety using dynamic computer simulations.

RESIDENTIAL RETROFIT

Side-to-side forces cause most earthquake damage. Bolting of the mudsill to the foundation and application of plywood to cripple walls are a few basic retrofit techniques which homeowners may apply to wood-framed residential structures to mitigate the effects of seismic activity. The City of San Leandro created guidelines for these procedures, as outlined in the following pamphlet. Public awareness and initiative are critical to the retrofit and preservation of existing building stock, and such efforts as those of the Association of Bay Area Governments are instrumental in providing informational resources to seismically active communities.

Wood Frame Structure

Most houses in North America are wood-framed structures. Wood is one of the best materials for earthquake-resistant construction since it is lightweight and more flexible than masonry. It is easy to work with and less expensive than steel, masonry, or concrete. In older homes the most significant weaknesses are the connection from the wood-framed walls to the foundation and the relatively weak "cripple-walls." (Cripple walls are the short wood walls that extend from the top of the foundation to the lowest floor level in houses that have raised floors.) Adding connections from the base of the wood-framed structure to the foundation is almost always an important part of a seismic retrofit. Bracing the cripple-walls to resist side-to-side forces is essential in houses with cripple walls; bracing is usually done with plywood. Oriented strand board (OSB) does not perform as consistently as plywood, and is not the favored choice of retrofit designers or installers.

Retrofit methods in older woodframe structures may consist of the following, and other methods not described here.

- The lowest plate rails of walls (usually called "mudsills" or "foundation sills" in North America) are bolted to a continuous foundation, or secured with rigid metal connectors bolted to the foundation so as to resist side-to-side forces.
- *Cripple walls* are braced with plywood.
- Selected vertical elements (typically the posts at the ends of plywood wall bracing panels) are connected to the foundation. These connections are intended to prevent the braced walls from rocking up and down when subjected to back-and-forth forces at the top of the braced walls, not to resist the wall or house "jumping" off the foundation (which almost never occurs).
- In two story buildings using "platform framing" (sometimes called "western" style construction, where walls are progressively erected upon the lower story's upper diaphragm, unlike "eastern" or *balloon framing*), the upper walls are connected to the lower walls with tension elements. In some cases, connections may be extended vertically to include retention of certain roof elements. This sort of strengthening is usually very costly with respect to the strength gained.
- Vertical posts are secured to the beams or other members they support. This is particularly important where loss of support would lead to collapse of a segment of a building. Connections from posts to beams cannot resist appreciable side-to-side forces; it is much more important to strengthen around the perimeter of a building (bracing the cripple-walls and supplementing foundation-to-wood-framing connections) than it is to reinforce post-to-beam connections.

Wooden framing is efficient when combined with masonry, if the structure is properly designed. In Turkey, the traditional houses (bagdadi) are made with this technology. In El Salvador, wood and bamboo are used for residential construction.

Reinforced and Unreinforced Masonry

In many parts of developing countries such as Pakistan, Iran and China, unreinforced or in some cases reinforced masonry is the predominantly form of structures for rural residential and dwelling. Masonry was also a common construction form in the early part of the 20th century, which implies that a substantial number of these at-risk masonry structures would have significant heritage value. Masonry walls that are not reinforced are especially hazardous. Such structures may be more appropriate for replacement than retrofit, but if the walls are the principal load bearing elements in structures of modest size they may be appropriately reinforced. It is especially important that floor and ceiling beams be securely attached to the walls. Additional vertical supports in the form of steel or reinforced concrete may be added.

In the western United States, much of what is seen as masonry is actually brick or stone veneer. Current construction rules dictate the amount of *tie–back* required, which consist of metal straps secured to vertical structural elements. These straps extend into mortar courses, securing the veneer to the primary structure. Older structures may not secure this sufficiently for seismic safety. A weakly secured veneer in a house interior (sometimes used to face a fireplace from floor to ceiling) can be especially dangerous to occupants. Older masonry chimneys are also dangerous if they have substantial vertical extension above the roof. These are prone to breakage at the roofline and may fall into the house in a single large piece. For retrofit, additional supports may be added; however, it is extremely expensive to strengthen an existing masonry chimney to conform with contemporary design standards. It is best to simply remove the extension and replace it with lighter materials, with special metal flue replacing the flue tile and a wood structure replacing the masonry. This may be matched against existing brickwork by using very thin veneer (similar to a tile, but with the appearance of a brick).

SEISMIC RESPONSE OF LANDFILL

Solid waste landfills can be affected by seismic activity. The tension in a landfill liner rises significantly during an earthquake, and can lead to stretching or tearing of the material. The top of the landfill may crack, and methane collection systems can be moved relative to the cover.

Increasing the depth of a waste column typically leads to a decrease in the ground acceleration felt at the surface of that landfill during an earthquake. The weight of waste is important in the analysis of landfill liner puncture and pipe crushing during an earthquake. Unstable rock under landfills such as limestone may yield during an earthquake, leading to a partial collapse of the fill. A major concern in this case would be the potential contamination of water sources that may be located below the landfill.

Modeling

In a pseudostatic slope stability analysis, the earthquake is modeled as a constant horizontal force. The fact that the earthquake force is modeled as a constant force acting in one direction, represents this model's major limitation. In a Newmark permanent deformation mathematical analysis, movement of a landfill occurs when a driving force on the landfill is greater than its resisting force. A shaking table laboratory test works to explore the strength characteristics at interfaces between different components of the landfill. Of primary concern is the contact between soil and landfill liner, as this is usually considered to be a weak point in the system. A dynamic centrifuge test works to model the landfill in a scaled-down form. Typically in this approach, a small simplified landfill is constructed in a test box. During this testing, earthquake loading is simulated on the sample landfill. A major application of this type of testing is to observe the behavior of landfill liners during ground shaking.

The uniqueness of many landfills makes it difficult to apply test results from one site to another. Measuring properties of landfills such as waste unit weight could lead to health concerns, as the handling of waste could be dangerous. Accessing this waste would likely involve damaging the cover liner, which could compromise the ability of the system to function properly. Landfills are systems which change over time, so periodic evaluations may be necessary.

REFERENCES

1. NZSEE Bulletin 39(2)-June 2006
2. ASCE-SEI 41
3. NZSEE 2006
4. Reitherman, Robert (2012). Earthquakes and Engineers: An International History. Reston, VA: ASCE Press. pp. 486–487. ISBN 9780784410714.
5. Moehle, J. (2000) State of Research on Seismic Retrofit
6. Filiatrault & Cherry (1986)
7. e.g. Kam & Pampanin (2008)- Selective weakening retrofit for RC frames
8. 1994 Building Publications - Status of the U.S. Precast Seismic Structural Systems (PRESSS) Program
9. Lowes & Moehle (1998) - ACI Structural Journal Vol 96(4) - pp 519-532
10. Experimental testing of external post-tensioning for retrofit of RC beam-column joint
11. VSL Repair/Strengthening Page
12. Clark Construction Group, LLC
13. Beres, A., Pessiki, S., White, R., and Gergely, P. (1996).
14. Implications of experimental on the seismic behaviour of gravity load designed RC beam-column connections. Earthquake Spectra, 12(2), 185-198.
15. Calvi, G. M., Moratti, M., and Pampanin, S. (2002). Relevance of beam-column damage and collapse in RC frame assessment. Journal of Earthquake Engineering, 6(1), 75-100.
16. Park, R. (2002). A Summary of Result of Simulated Seismic Load Tests on Reinforced Concrete Beam-Column Joints, Beams and Columns with Substandard Reinforcing Details. Journal of Earthquake Engineering, 6(2), 147-174.
17. Park R, Billings IJ, Clifton GC, Cousins J, Filiatrault A, Jennings DN, et al. The Hyogo-ken Nanbu Earthquake of 17 January 1995. Bull of New Zealand Soc of earthquake Eng. 1995;28(1):1 -99.

18. Holmes WT, Somers P. Northridge earthquake Reconnaissance Report. Supplement C, vol. 2. earthquake Spectra. 1996(11):1-278.
19. Pampanin, S., Bolognini, D., Pavese, A. (2007) Performance-based Seismic Retrofit Strategy for Existing Reinforced Concrete Frame Systems using FRP composites. ASCE Journal of Composites for Construction, 11(2), pp. 211-226.
20. A. Ghobarah and A. Said. 2002. Shear strengthening of beam-column joints. Engineering Structures, Vol. 24, No. 7, pp. 881- 888.
21. Ghobarah and A. Said 2001 Seismic rehabilitation of beam-column joints using FRP laminates. Journal of Earthquake Engineering, Vol. 5, No. 1, pp. 113-129.
22. Selective weakening and post-tensioning for seismic retrofit of RC beam-column joint
23. Bertero VV, Anderson JC & Krawinkler H. Performance of steel building structures during the Northridge earthquake. Report No UCB/EERC-94/09. Berkeley, California: Earthquake Engineering Research Center, University of California at Berkeley. 1994.
24. Civjan SA, Engelhardt MD and Gross JD (2000). Retrofit of pre-Northridge Moment Resisting Connections. ASCE J.o.Structural Engineering Vol 126(4) 445-452
25. FEMA 350, July 2000. Recommended Seismic Design Criteria for New Steel Moment-Frame Buildings.1.3, pages 1-3 through 1-11.
26. Thusyanthan, N.I.; Madabhushi, S.P.G.; Singh, S. (2007) "Tension in Geomembranes on Landfill slopes Under Static and Earthquake Loading - Centrifuge Study" Geotextiles and Geomembranes V. 25: 78-95.
27. Matasovic, N.; Kavazanjian Jr., E. (2006) "Seismic Response of a Composite Landfill Cover" Journal of Geotechnical and Geoenvironmental Engineering V. 132(4): 448-455.
28. Bray, J.D.; Augello, A.J.; Leonards, G.A.; Repetto, P.C.; Byrne, R.J. (1995) "Seismic Stability Procedures for Solid-Waste Landfills" Journal of Geotechnical Engineering V. 121(2): 139-151.
29. Choudhury, D.;Savoikar, P. (2009) "Simplified Method to Characterize Municipal Solid Waste Properties Under Seismic Conditions" Waste Management V. 29(2):924-933.
30. Krinitzsky, E.L.;Hynes, M.E;Franklin, A.G (1997) "Earthquake Safety Evaluation of Sanitary Landfills" Engineering Geology V. 46(2): 143-156.

Index

V